Surveys of Modern Mathematics

www.intlpress.com/smm

SMM 1: *Analytic Methods in Algebraic Geometry*
by Jean-Pierre Demailly

SMM 2: *Lie Theory and Representation Theory*
Naihong Hu, Bin Shu and Jianpan Wang, eds.

SMM 3: *Application of Elementary Differential Geometry to Influence Analysis*
Yat-Sun Poon and Wai-Yin Poon, eds.

SMM 4: *An Introduction to Rota-Baxter Algebra*
by Li Guo

SMM 6: *Open Problems and Surveys of Contemporary Mathematics*
Lizhen Ji, Yat-Sun Poon and Shing-Tung Yau, eds.

SMM 7: *Introductory Lectures on Manifold Topology: Signposts*
by Thomas Farrell and Yang Su

SMM 8: *Lie-Bäcklund-Darboux Transformations*
by Y. Charles Li and Artyom Yurov

SMM 9: *Compressible Flow and Euler's Equations*
by Demetrios Christodoulou and Shuang Miao

SMM 10: *Finite Groups: An Introduction*
by Jean-Pierre Serre

SMM 11: *Topics in Differential Geometry:*
Selected papers & lecture notes of Shiing-Shen Chern
edited by Shiu-Yuen Cheng and Lizhen Ji

SMM 12: *Differential Geometry & Integral Geometry:*
Selected papers & lecture notes of Shiing-Shen Chern
edited by Shiu-Yuen Cheng and Lizhen Ji

SMM 13: *Gauss–Manin Connection in Disguise: Calabi–Yau Modular Forms*
by Hossein Movasati

Surveys of Modern Mathematics

www.intlpress.com/smm

Surveys of Modern Mathematics
Volume 13

Gauss–Manin Connection in Disguise:

Calabi–Yau Modular Forms

Hossein Movasati

International Press
www.intlpress.com

HIGHER EDUCATION PRESS

Surveys of Modern Mathematics, Volume 13

Gauss–Manin Connection in Disguise: Calabi–Yau Modular Forms

by Hossein Movasati)

ISBN: 978-1-57146-343-2

Printed in the United States of America.

9 8 7 6 5 4 3 2 1

To my parents Rogayeh and Ali,
and my family Sara and Omid

Preface

The guiding principle in this monograph is to develop a theory of (Calabi-Yau) modular forms parallel to the classical theory of (elliptic) modular forms. It is originated from many period manipulations of the B-model Calabi-Yau variety of mirror symmetry in Topological String Theory and the earlier works of the author in which the theory of (quasi) modular forms is introduced using a larger moduli space of elliptic curves, and the Ramanujan differential equations between Eisenstein series have been derived from the corresponding Gauss-Manin connection. We have in mind an audience with a basic knowledge of Complex Analysis, Differential Equations, Algebraic Topology and Algebraic Geometry. Although the text is purely mathematical and no background in String Theory is required, some of our computations are inspired by mirror symmetry, and so the reader who wishes to explore the motivations, must go to the original Physics literature. The text is mainly written for two primary target audiences: experts in classical modular and automorphic forms who wish to understand the q-expansions of physicists derived from Calabi-Yau threefolds, and mathematicians in enumerative Algebraic Geometry who want to understand how mirror symmetry counts rational curves in compact Calabi-Yau threefolds. Experts in modular forms are warned that they will not find so much Number Theory in the present text, as this new theory of modular forms lives its infancy, and yet many problems of complex analysis nature are open.We have still a long way to deal with more arithmetic oriented questions. For our purpose we have chosen a particular class of such q-expansions arising from the periods of a Calabi-Yau threefold called mirror quintic, and in general, periods which satisfy fourth-order differential equations. The applications of classical modular forms are huge and we are guided by the fact that this new type of modular forms might have similar applications in the near future, apart from counting rational curves and Gromov-Witten invariants. The main goal is to describe in detail many analogies and differences between classical modular forms and those treated here. The present text is a complement to the available books on the mathematical aspects of mirror symmetry such as "Mirror Symmetry and Algebraic Geometry" of D. A. Cox and S. Katz and "Mirror Symmetry" of C. Voisin. We hope that our text makes a part of mirror symmetry, which is relevant to Number Theory, more accessible to mathematicians.

Hossein Movasati
December 2016
Rio de Janeiro, RJ, Brazil

Acknowledgements

The present text was written during the years 2010 through 2014. First of all, I would like to thank Pierre Deligne for all his comments on the origin [Mov12b] of the present text. My sincere thanks go to Étienne Ghys who drew my attention to the historical aspects of Ramanujan differential equations, that is, the contribution of Darboux and Halphen which is mainly neglected in Number Theory. My sincere thanks go to Charles Doran, Stefan Reiter, Duco van Straten, Don Zagier, Conan Leung, Bong Lian and Babak Haghighat for useful conversations and their interest on the topic of the present text. I would also like to thank mathematics institutes Instituto de Matemática Pura e Aplicada (IMPA) in Rio de Janeiro (my home institute), Max-Planck Institute for Mathematics (MPIM) in Bonn, Institute for Physics and Mathematics (IPM) in Tehran, Institute for Mathematical Sciences (IMS) in Hong Kong, Center of Advanced Study in Theoretical Sciences (CASTS) in Taipei and Center of Mathematical Sciences and Applications (CMSA) at Harvard University for providing excellent research ambient during the preparation of the present text. Final versions of the present text are written when the author was spending a sabbatical year at Harvard University. Here, I would like to thank Shing-Tung Yau for the invitation and for his interest and support. My sincere thanks go to Murad Alim and Emmanuel Scheidegger for many useful conversations in mathematical aspects of Topological String Theory and mirror symmetry. I would like to thank Cumrun Vafa for his generous emails helping me to understand the mathematical content of anomaly equations. During the preparation of Appendix C we enjoyed discussions with many people. We would like to thank M. Belolipetsky for taking our attention to the works on thin groups. Thanks also go to Peter Sarnak, Fritz Beukers and Wadim Zudilin for comments on the first draft of this appendix. The second author of this appendix would like to thank CNPq-Brazil for financial support and IMPA for its lovely research ambient.

Contents

Frequently used notations

$(\mathbb{C}^n, 0)$	A small neighborhood of 0 in $\mathbb{C}^n$.
$\mathsf{k}, \bar{\mathsf{k}}$	A field of characteristic zero and its algebraic closure.
M^{tr}	The transpose of a matrix M. We also write $M = [M_{ij}]$, where M_{ij} is the (i, j) entry of M. The indices i and j always count the rows and columns, respectively.
$V^\vee$	The dual of an R-module V, where R is usually the ring $\mathbb{Z}$ or the field k. We always write a basis of a free R-module of rank r as a $r \times 1$ matrix. For a basis δ of V and α of $V^\vee$ we denote by $$[\delta, \alpha^{\mathsf{tr}}] := [\alpha_j(\delta_i)]_{i,j}$$ the corresponding $r \times r$ matrix.
d	The differential operator or a natural number, being clear in the text which one we mean.
X	A mirror quintic Calabi-Yau threefold, mirror quintic for short, or an elliptic curve defined over the field k, being clear in the text which we mean, §3.1. We will also use X as one of the Yamaguchi-Yau variables in §2.17. Another usage of X is as a fundamental system of a linear differential equation, §7.4.
$X(\mathsf{k})$	The set of k-rational points of X defined over the field k. In particular for $\mathsf{k} \subset \mathbb{C}$, $X(\mathbb{C})$ is the underlying complex manifold of X. Sometimes, for simplicity we write $X = X(\mathbb{C})$, being clear in the context that X is a complex manifold.
ω, η	A differential 3-form on X. In many cases it is a holomorphic $(3,0)$-form.
$H_3(X, \mathbb{Z})$	The third homology of $X = X(\mathbb{C})$.
δ_i, $i = 1,2,3,4$	A symplectic basis of $H_3(X, \mathbb{Z})$.
T	Moduli of enhanced mirror quintics over $\bar{\mathsf{k}}$, §2.3.
S	A moduli of mirror quintics over $\bar{\mathsf{k}}$, §3.2.
$\mathscr{O}_{\mathsf{T}}, \mathscr{O}_{\mathsf{S}}$	The k-algebra of regular functions on T and S, $\mathsf{T} = \mathrm{Spec}(\mathscr{O}_{\mathsf{T}})$, etc.

$t_0,\ i=0,\dots,9$	Regular functions in T, §3.2, §3.6.
R	Vector fields on T.
Ω^i_{T}	The $\mathscr{O}_{\mathsf{T}}$-module of differential i-forms on T.
$\mathsf{X}\to\mathsf{T},\ \mathsf{X}/\mathsf{T}$	The universal family of enhanced mirror quintics. A single variety is denoted by X, §3.6.
$\mathsf{X}_t,\ t\in\mathsf{T}$	A fiber of $\mathsf{X}\to\mathsf{T}$.
$H^*_{\mathrm{dR}}(X)$	Algebraic de Rham cohomology, §3.6.
$H^*_{\mathrm{dR}}(\mathsf{X}/\mathsf{T})$	Relative algebraic de Rham cohomology, §3.6.
$F^*H^*_{\mathrm{dR}}(X)$	The pieces of Hodge filtration, §3.6.
$F^*H^*_{\mathrm{dR}}(\mathsf{X}/\mathsf{T})$	The pieces of Hodge filtration, §3.6.
$\alpha_i,\ i=1,\dots,4$	A basis of $H_3(X)$ or $H^3(\mathsf{X}/\mathsf{T})$, §2.3.
$\langle\cdot,\cdot\rangle$	The intersection form in de Rham cohomology and singular homology, §3.4, §4.1.
$\delta^{\mathsf{pd}}\in H^3_{\mathrm{dR}}(X)$	The Poincaré dual of $\delta\in H_3(X,\mathbb{Z})$, §4.1.
Φ,Ψ	

$$\Phi:=\begin{pmatrix}0&0&0&1\\0&0&1&0\\0&-1&0&0\\-1&0&0&0\end{pmatrix},\quad \Psi:=\begin{pmatrix}0&0&1&0\\0&0&0&1\\-1&0&0&0\\0&-1&0&0\end{pmatrix} \tag{0.1}$$

G,g	An algebraic group and its element, §3.10.
$\mathrm{Lie}(\mathsf{G}),\ \mathfrak{g}$	The Lie algebra of G and its element, §3.10.
$\mathsf{g}_i,\ i=1,\dots,6$	A parametrization of $\mathsf{g}\in\mathsf{G}$, §3.10.
∇	Gauss-Manin connection, §3.3, §3.8.
A	The Gauss-Manin connection matrix, §3.3, §3.8.
Y	Yukawa coupling, §2.3.
E_k	Eisenstein series of weight k, §2.5
θ	The derivation $z\frac{\partial}{\partial z}$, §2.7, or a special period matrix, §4.12.
M_0,M_1	Monodromy matrices, §2.7.
Γ	The monodromy group, §2.7
$\mathbb{H}$	Upper half plane or the monodromy covering, §4.6.
τ	The canonical coordinate of the upper half plane or a special period matrix, §4.2.
$\tau_i,\ i=0,1,2,3$	Meromorphic functions on $\mathbb{H}$, §2.10, §10.2.
$\theta_i,\ i=0,1,2,3$	Meromorphic functions on $\mathbb{H}$, §4.12, §10.2.
$\mathsf{F}^{\mathrm{alg}}_g,\mathsf{F}^{\mathrm{non}}_g,\mathsf{F}^{\mathrm{hol}}_g$	Genus g topological string partition functions, §2.13.
L	A fourth-order linear differential equations, §2.15.
ψ_0,ψ_1	Holomorphic and logarithmic solutions of $L=0$, §2.7, §7.1.
x_{ij}	Particular solutions of $L=0$ and their derivatives, §2.7, §7.4. We denote by $\mathsf{X}:=[x_{ij}]$ the corresponding matrix.
y_{ij}	General solutions of $L=0$ and their derivatives, §4.15.
$\mathcal{A}$	A linear system attached to $L=0$, §7.3.
$a_i(z)$	Coefficients of L, §2.15.
u_i	Expressions in terms of the solutions of L, §2.15.
X,U,V_1,V_2,V_3	Yamaguchi-Yau variables, §2.17, §6.1, §8.4.

P	Period map or period matrix, §4.1.
Π	Generalized period domain, §4.4.
$\mathrm{F}, \mathrm{G}, E$	Intersection matrices, §7.4.
χ	The Euler number of the underlying Calabi-Yau threefolds of $L = 0$, §7.3.
h^{21}	The $(2,1)$ Hodge number of the underlying Calabi-Yau threefolds of $L = 0$. The text mainly deals with $h^{21} = 1$.

Online supplemental items

Many arguments and proofs of the present text rely on heavy computer computations. For this purpose, we have used Singular, [GPS01], a computer programming language for polynomial computations. Throughout the text I have used notations in the form [Supp Item x], where x=1, 2, 3, etc. Each of these notations refers to a supplemental item, such as computer data or code, etc., that can be accessed online from my web page:

http://w3.impa.br/∼hossein/singular/GMCD-MQCY3-SuppItems.html

The web page serves as a hot-linked index to all of this book's online supplemental items. These supplemental items are mainly for two purposes. First, they are mainly for the reader who does not want to program by her or himself and wants to check the statements using our computer codes. Second, we only present a small amount of computer data in the text, and the reader can use the supplemental item in order to access to more data, for instance more coefficients of a q-expansion of a series. We have also written a library in Singular [Supp Item 1] which has been useful for our computations.

Chapter 1
Introduction

Topological String Theory and in particular mirror symmetry have provided us with many q-expansions which at first glance look like classical modular forms. From this we have mainly in mind those calculated by Candelas et al. in [CdlOGP91] and Bershadsky et al. in [BCOV93, BCOV94]. In 1991 Candelas, de la Ossa, Green and Parkes in [CdlOGP91] calculated in the framework of mirror symmetry a generating function, called the Yukawa coupling, which predicts the number of rational curves of a fixed degree in a generic quintic threefold. The pair of a quintic threefold and its mirror $(\hat{X}, X)$ is just an example of the so-called mirror Calabi-Yau threefolds. The miracle of mirror symmetry is that some simple period computations in the mirror X (the B-model of mirror symmetry) gives us predictions on the number of curves in $\hat{X}$ (the A-model of mirror symmetry). The first manifestation of mirror symmetry is in the Hodge diamond of the mirror pairs. The mirror quintic Calabi-Yau threefold X has the Hodge diamond

$$\begin{array}{ccccccc} & & & 1 & & & \\ & & 0 & & 0 & & \\ & 0 & & 101 & & 0 & \\ 1 & & 1 & & 1 & & 1 \\ & 0 & & 101 & & 0 & \\ & & 0 & & 0 & & \\ & & & 1 & & & \end{array}$$

whereas after 90° rotation we get the Hodge diamond of $\hat{X}$, a smooth quintic threefold. In some interesting cases of non-compact mirror threefolds (see [ABKv2] and the references therein) and elliptic curves (see [Dij95]), the computed q-expansions are in fact (quasi-) modular forms, however, for compact Calabi-Yau threefolds such q-expansions are essentially different from all previously well-known special functions. Since 1991 there were many efforts, mainly among experts in Number Theory and modular forms, to relate such q-expansions, and in particular the Yukawa coupling, to classical modular or quasi-modular forms and in this way find modular properties of such functions. However, there have been not so much success in this direction. Classical modular forms build a finitely generated algebra with finite dimensional pieces and this simple fact leads us to many applications of modular forms ranging from number theory, enumerative Algebraic Geometry, representation theory of groups and Lie algebras to non-commutative geometry, see for instance [Zag08, Gan06] and the references therein. This urges us to look for such algebras and their applications in the case of compact Calabi-Yau threefolds. The indication for the existence of such algebras for Calabi-Yau threefolds has been partially appeared in the works of Bershadsky, Cecotti, Ooguri and Vafa [BCOV93, BCOV94]. In these articles the authors introduce the so-called BCOV holomorphic anomaly equation which is a differential equation satisfied by topological partition functions F_g. The function F_g gives us the generating function for the number of genus g and degree d curves in the A-model Calabi-Yau variety. In order to analyze the internal structure of F_g's, they used Feynman propagators which play the role of a basis of some algebra which is big enough to solve the BCOV anomaly equation. Following this idea S. Hosono in [Hos08], formulated BCOV rings which are in general infinitely generated. The first indication that there must be a finitely generated algebra which solves the BCOV anomaly equation, was made by Yamaguchi-Yau in [YY04] in the one-dimensional moduli of Calabi-Yau varieties and later by Alim-Länge in [AL07] for the multi dimensional case. They showed that Feynman propagators and their derivatives can be written in terms of finite number of generators.

Despite all the developments above, there is no literature in which we find a mathematical theory which develops the q-expansions of Physics in the same style as of modular forms, available in many classical books such as [Lan95a, Miy89, DS05, Zag08]; to state just few books among many dedicated to modular forms. The main reason for this is that the Griffiths period domain for compact Calabi-Yau threefolds is not a Hermitian symmetric domain and so it does not enjoy any automorphic function theory, see [Gri70]. The main objective of the present text is to describe a modular form theory attached to Calabi-Yau varieties, which clearly shows the analogies and differences with well-known modular form theories. We believe that this is indispensable for the future applications of all such q-expansions both in Physics and Number Theory. To do this we will stick to the most well-known compact Calabi-Yau threefold used in the B-model of mirror symmetry, that is, the mirror to a generic quintic threefold, for definition see §3.1. We call it a mirror quintic Calabi-Yau threefold, mirror quintic for short. Once this is done, it would not be difficult to develop similar theories for Calabi-Yau varieties with one and multi dimensional moduli spaces, see [AMSY14]. Attached to mirror quintics, we describe 7 convergent series $t_i(q), i = 0,1,2,\dots,6$ and we explain how the algebra $\mathbb{C}[t_i(q),\ \ i = 0,1,2,\dots,6]$ can be considered as a genuine generalization of the algebra of quasi-modular forms generated by the classical Eisenstein series E_{2i}, $i = 1,2,3$. In fact, we construct three different notions of such t_i's: holomorphic, anti-holomorphic and algebraic versions. The holomorphic and anti-holomorphic versions were already used in [BCOV93, BCOV94], however, the algebraic version, which turns such q-expansions to be defined over arbitrary fields, are quite new and can be considered as one of the contributions of our text. For the algebraic version, what we have first constructed, are seven affine coordinates t_i, $i = 0,1,2,\dots,6$

for the moduli of mirror quintics enhanced with certain elements in their third de Rham cohomologies. The main object for the related computations is not actually the mirror quintic itself, but its Picard-Fuchs equation L, which is a fourth-order linear differential equation. Almkvist, Enckevort, van Straten and Zudilin in [GAZ10] took the job of classifying such linear differential equations and by the time of writing this text, they were able to identify more than 400 of them. They called them Calabi-Yau equations. For most of them they were able to identify the corresponding family of Calabi-Yau varieties. Our computations go smoothly for all Calabi-Yau equations. Attached to such an equation we introduce a differential field of transcendental degree 7 which in the mirror quintic case it is the quotient field of $\mathbb{C}[t_i(q),\ i=0,1,\ldots,6]$. In such a generality, we have only holomorphic and anti-holomorphic version of such a field which are related to each other by the notion of holomorphic limit. The algebraic version of such a field, and more importantly, a finitely generated algebra which can be considered as a generalization of the algebra of modular forms, requires a detailed study of the corresponding family of Calabi-Yau varieties and their degenerations, something which cannot be seen directly from the Picard-Fuchs equation.

As a secondary objective, we aim to present the mathematical content of the Physics literature on topological partition functions and BCOV anomaly equations, see [BCOV93, BCOV94] and the later developments [YY04, HKQ09, Hos08]. This literature in Physics produces many predictions and conjectures for mathematicians, and in particular algebraic geometers, and it would important to write them down in a mathematical paper avoiding many physical theories which requires less mathematical rigor and more physical intuition. From this point of view, our text is mainly expository, however, our exposition gives the proof of many statements which are not rigorously established in the Physics literature.

The text is mainly written for number theorists who want to know what are the generalizations of modular forms which arise in Physics literature, and for algebraic geometers who want to know how physicists using mirror symmetry compute the number of curves in Calabi-Yau varieties, for instance in a generic quintic threefold. The text is supposed to be elementary and a basic knowledge of Complex Analysis, Differential Equations, Algebraic Topology and Algebraic Geometry would be enough to follow it. For computations throughout the text the reader might need some ability to work with a mathematical software for dealing with problems related to commutative algebra and formal power series.

1.1 What is Gauss-Manin connection in disguise?

In [Mov08a, Mov12b], the author realized that the Ramanujan relations between Eisenstein series can be computed using the Gauss-Manin connection of families of elliptic curves and without knowing about Eisenstein series, see §2.2. Then the author built up the whole theory of quasi-modular forms, see [Dij95, KZ95], based on this geometric observation. Later, Pierre Deligne in a private communication, see

[Del09], called this the *Gauss-Manin connection in disguise*. The main contribution of the present text is to built a modular form theory attached to mirror quintics in the same style, that is, to start from a purely algebraic geometric context, construct the generalization of the Ramanujan differential equation and built up the whole theory upon this.

The present text is a part of a project called "Gauss-Manin Connection in Disguise". In the first text [AMSY14] in this series, jointly written with M. Alim, E. Scheidegger and S. T. Yau, we presented an algebro-geometric framework for a modular form theory attached to non-rigid compact Calabi-Yau threefolds. Our approach is based on a moduli T of enhanced projective varieties introduced in [Mov13], that is, a projective variety X with a fixed topological data and enhanced with a basis of the algebraic de Rham cohomologies $H^*_{\rm dR}(X)$, compatible with the Hodge filtration and with a constant cup product structure. We argued that the algebra $\mathscr{O}_\mathsf{T}$ of regular functions on T contains genus g topological string partitions functions, and these encode the Gromov-Witten invariants of Calabi-Yau threefolds. The emphasis in [AMSY14] was mainly on geometric aspects. There, we did not push forward the analogies and differences between the algebra $\mathscr{O}_\mathsf{T}$ and the algebra of quasi-modular forms. No automorphic function theory and Fourier (or q-) expansions were discussed in the mentioned text. The first example in which we discussed both topics, has been appeared in the author's earlier texts on elliptic curves and quasi-modular forms, see [Mov12b, Mov08a, Mov08b]. In the present text we fill this gap by working out the case of mirror quintics and in general Calabi-Yau threefolds with a one-dimensional complex moduli space. This will give us a non-classical theory of modular forms. Since for modular forms, and in general automorphic forms, we have a huge amount of articles and many books, I think, this justifies why the material of the present text deserves a separate presentation. The knowledge of [AMSY14] is not necessary for reading the present text, as most of the material presented here were written before that, however, following the same story in the case of elliptic curves and quasi-modular forms as in [Mov12b] will help the reader to understand the motivations behind many computations throughout the present text.

1.2 Why mirror quintic Calabi-Yau threefold?

During many years that I have been working on various aspects of mirror quintics, I have been asked why one should put so much effort on just one example of Calabi-Yau varieties. Here, I would like to answer this question. First of all, the question does not make sense if one remembers that an elliptic curve is also an example of a Calabi-Yau variety, however, the arithmetic of elliptic curves is a huge area of research and, for instance, Birch and Swinnerton conjecture on L functions of elliptic curves is one of the millennium conjectures. In dimension two we have now a good amount of research on arithmetic of K3 surfaces, however, we do not know too much about the arithmetic of non-rigid compact Calabi-Yau threefolds, and in particular mirror quintics. In this case even many topological and complex analysis questions, which are trivialities for elliptic curves, turn out to be hard problems, see §11. Since my acquaintance with the arithmetic modularity, see for instance Diamond-Shurman's book [DS05], I have always believed that modular forms are responsible for the modularity of a limited number of varieties in Arithmetic Algebraic Geometry. All the attempts to find an arithmetic modularity for mirror quintics have failed, see for instance Yui's expository article [Yui13], and this might be an indication that maybe such varieties need a new kind of modular forms. In the present text I suggest one, however, we are still far from arithmetic issues. I would also like to remind that the mirror symmetry is still a mysterious statement in the case of quintic threefolds. Mirror symmetry predicts that the manipulations of periods for mirror quintics, actually compute the virtual number of genus g and degree d curves in a generic quintic. Mathematical verification of these predictions are done for $g = 0, 1$ and for higher genus it is still an open problem. Finally, I would like to point out Clemens' conjecture, which roughly speaking states that, what mirror symmetry attempts to count in genus zero is actually a finite number.

1.3 How to read the text?

In §2 we give a summary of the results and computations obtained throughout the text. The reader may use this as a gateway to the rest of the text. It is specially for the reader who is not interested in details and wants to have a flavor of what is going on. This section is also intended to be the announcement of the new results, objects and computations of the present text. The expository and preparatory materials are scattered throughout the other sections. For the details of the material presented in §2 we refer to the corresponding sections. The content of §3, §4, §5, §6 is mainly for mirror quintics, whereas the content of §7, §8, §9 is for a big class of fourth-order linear differential equations. These two sets of chapters can be read independently. The content of §10, §11 and Appendices A, B, C, D are independent of the rest of the text, however, for some notations we have referred to the main body of the text.

1.4 Why differential Calabi-Yau modular form?

During the years I was working on this monograph, I had some difficulty choosing a proper name for the modular form theory developed in the present text. Since the Physics literature only deals with examples of such modular forms, it would not be reasonable to use names such as Yukawa couplings, topological string partition functions, propagators etc. In [Mov15] and the earlier drafts of the present text, I chose the name modular-type function which does not seem to be appropriate. An expert in Number Theory was against the usage of the term modular form, as there is no upper half plane in our theory. In my earlier work [Mov08a], which was written without any awareness of Physics literature and in particular [Dij95, KZ95], I used the term differential modular form instead of quasi-modular form. This naming was originated from the works of Buium, see for instance [Bui00]. Since the adjective differential represents the differential structure of this new type of modular form theories, it is reasonable to use it instead of quasi. Finally, since we are dealing with Calabi-Yau varieties and their moduli spaces, I decided to use the name differential Calabi-Yau modular form for the generalization of quasi-modular forms in the present text. The term Calabi-Yau modular form will be reserved for a generalization of classical modular forms that will be briefly discussed in §10.

Chapter 2
Summary of results and computations

In this chapter we give a summary of the material of the present text. For more details on each topic we refer to the corresponding section. All the story presented in this text has its origin in the classical context of elliptic curves and modular forms, see [Mov12b]. Our approach to modular forms is purely geometric and it is through differential equations, which might have some novelties even for experts in modular forms. Therefore, we will also review some results and computations in [Mov12b]. This might help the reader to understand our approach to differential Calabi-Yau modular forms attached to mirror quintic Calabi-Yau threefolds, mirror quintic for short. We are not going to produce different notations in order to distinguish the case of elliptic curves and mirror quintics. Hopefully, this will not produce any confusion.

2.1 Mirror quintic Calabi-Yau threefolds

Instead of an elliptic curve we are going to work with a mirror quintic X. The reader can actually skip the definition presented in §3.1 and bear the following data in mind. It is a smooth projective threefold and its third algebraic de Rham cohomology $H^3_{\rm dR}(X)$ is of dimension 4 and it carries the so-called Hodge filtration:

$$\{0\} = F^4 \subset F^3 \subset F^2 \subset F^1 \subset F^0 = H^3_{\rm dR}(X), \ \ \dim_{\mathbb{C}}(F^i) = 4 - i. \tag{2.1}$$

For $\omega, \alpha \in H^3_{\rm dR}(X)$ let

$$\langle \omega, \alpha \rangle := {\rm Tr}(\omega \cup \alpha)$$

be the intersection form. The trace map Tr is defined for an arbitrary field k, however, it is useful to carry in mind that for $\mathsf{k} = \mathbb{C}$ it is just

$${\rm Tr}(\alpha) := \frac{1}{(2\pi i)^3} \int_{X(\mathbb{C})} \alpha, \ \ \alpha \in H^6_{\rm dR}(X).$$

Recall that the first de Rham cohomology $H^1_{\mathrm{dR}}(X)$ of an elliptic curve X over k is a two-dimensional k-vector space, and it carries the Hodge filtration $\{0\} = F^2 \subset F^1 \subset F^0 = H^1_{\mathrm{dR}}(X)$, where the non-trivial piece F^1 is generated by a regular differential 1-form in X. Further, $\mathrm{Tr} : H^2_{\mathrm{dR}}(X) \cong \mathsf{k}$ is given by $\mathrm{Tr}(\alpha) := \frac{1}{2\pi i}\int_{X(\mathbb{C})} \alpha$.

2.2 Ramanujan differential equation

Let T be the moduli of the pairs $(E, [\alpha, \omega])$, where E is an elliptic curve over k and $\omega \in H^1_{\mathrm{dR}}(E)$, $\alpha \in F^1$ (α is a regular differential form on E) such that $\langle \alpha, \omega \rangle = 1$.

Theorem 1 *([Mov12b], Proposition 4.1, 5.4) There is a unique vector field* R *in* T*, such that the Gauss-Manin connection of the universal family of elliptic curves over* T *composed with the vector field* R*, namely* ∇_{R}*, satisfies*

$$\nabla_{\mathsf{R}}(\alpha) = -\omega,\ \nabla_{\mathsf{R}}(\omega) = 0. \tag{2.2}$$

In fact, T *has a canonical structure of an affine variety over* $\mathbb{Q}$

$$\mathsf{T} := \mathrm{Spec}(\mathbb{Q}[t_1, t_2, t_3, \frac{1}{27t_3^2 - t_2^3}])$$

and we have the corresponding universal family of elliptic curves

$$y^2 = 4(x-t_1)^3 - t_2(x-t_1) - t_3,\ \ \alpha = [\frac{dx}{y}],\ \omega = [\frac{xdx}{y}]. \tag{2.3}$$

The Gauss-Manin connection of the above family written in the basis α*,* ω *is given by*

$$\nabla \begin{pmatrix} \alpha \\ \omega \end{pmatrix} = \mathsf{A} \begin{pmatrix} \alpha \\ \omega \end{pmatrix} \tag{2.4}$$

where

$$\mathsf{A} = \frac{1}{\Delta}\begin{pmatrix} -\frac{3}{2}t_1\alpha - \frac{1}{12}d\Delta & \frac{3}{2}\alpha \\ \Delta dt_1 - \frac{1}{6}t_1 d\Delta - (\frac{3}{2}t_1^2 + \frac{1}{8}t_2)\alpha & \frac{3}{2}t_1\alpha + \frac{1}{12}d\Delta \end{pmatrix}, \tag{2.5}$$

$$\Delta = 27t_3^2 - t_2^3,\ \alpha = 3t_3dt_2 - 2t_2dt_3.$$

The vector field R *in the coordinates* t_1, t_2, t_3 *is given by*

$$\mathsf{R} = (t_1^2 - \frac{1}{12}t_2)\frac{\partial}{\partial t_1} + (4t_1t_2 - 6t_3)\frac{\partial}{\partial t_2} + (6t_1t_3 - \frac{1}{3}t_2^2)\frac{\partial}{\partial t_3}. \tag{2.6}$$

The vector field (2.36) written as an ordinary differential equation is known as Ramanujan relations between Eisenstein series. Theorem 1 gives us the differential equation of modular forms without constructing them. This is our starting point for generalizing modular forms.

2.3 Modular vector fields

We work with a moduli of enhanced mirror quintics and compute a modular differential equation. Let T be the moduli of pairs $(X,[\alpha_1,\alpha_2,\alpha_3,\alpha_4])$, where X is a mirror quintic Calabi-Yau threefold and

$$\alpha_i \in F^{4-i}/F^{5-i}, \quad i=1,2,3,4,$$

$$[\langle \alpha_i,\alpha_j\rangle] = \Phi.$$

Here, $H^3_{\mathrm{dR}}(X)$ is the third algebraic de Rham cohomology of X, F^i is the i-th piece of the Hodge filtration of $H^3_{\mathrm{dR}}(X)$, $\langle\cdot,\cdot\rangle$ is the intersection form in $H^3_{\mathrm{dR}}(X)$ and Φ is given by (0.1). In §3 we construct the universal family $\mathsf{X}\to\mathsf{T}$ together with global sections $\alpha_i,\ i=1,2,3,4$ of the relative algebraic de Rham cohomology $H^3(\mathsf{X}/\mathsf{T})$. Let

$$\nabla : H^3_{\mathrm{dR}}(\mathsf{X}/\mathsf{T}) \to \Omega^1_{\mathsf{T}} \otimes_{\mathscr{O}_{\mathsf{T}}} H^3_{\mathrm{dR}}(\mathsf{X}/\mathsf{T})$$

be the algebraic Gauss-Manin connection on $H^3(\mathsf{X}/\mathsf{T})$. The basic idea behind all the computations in the next sections lies in the following generalization of Theorem 1.

Theorem 2 *There is a unique vector field R_0 in T such the Gauss-Manin connection of the universal family of mirror quintic Calabi-Yau threefolds over T composed with the vector field R_0, namely ∇_{R_0}, satisfies*

$$\begin{aligned}\nabla_{\mathsf{R}_0}(\alpha_1) &= \alpha_2,\\ \nabla_{\mathsf{R}_0}(\alpha_2) &= \mathsf{Y}\alpha_3,\\ \nabla_{\mathsf{R}_0}(\alpha_3) &= -\alpha_4,\\ \nabla_{\mathsf{R}_0}(\alpha_4) &= 0\end{aligned} \tag{2.7}$$

for some regular function Y in T. In fact,

$$\mathsf{T} := \mathrm{Spec}\Big(\mathbb{Q}\Big[t_0,t_1,\ldots,t_6,\frac{1}{t_4t_5(t_4-t_0^5)}\Big]\Big), \tag{2.8}$$

the vector field R_0 as an ordinary differential equation is given by

$$\mathsf{R}_0: \begin{cases} \dot t_0 = \frac{1}{t_5}(6\cdot 5^4t_0^5 + t_0t_3 - 5^4t_4),\\ \dot t_1 = \frac{1}{t_5}(-5^8t_0^6 + 5^5t_0^4t_1 + 5^8t_0t_4 + t_1t_3),\\ \dot t_2 = \frac{1}{t_5}(-3\cdot 5^9t_0^7 - 5^4t_0^5t_1 + 2\cdot 5^5t_0^4t_2 + 3\cdot 5^9t_0^2t_4 + 5^4t_1t_4 + 2t_2t_3),\\ \dot t_3 = \frac{1}{t_5}(-5^{10}t_0^8 - 5^4t_0^5t_2 + 3\cdot 5^5t_0^4t_3 + 5^{10}t_0^3t_4 + 5^4t_2t_4 + 3t_3^2),\\ \dot t_4 = \frac{1}{t_5}(5^6t_0^4t_4 + 5t_3t_4),\\ \dot t_5 = \frac{1}{t_5}(-5^4t_0^5t_6 + 3\cdot 5^5t_0^4t_5 + 2t_3t_5 + 5^4t_4t_6),\\ \dot t_6 = \frac{1}{t_5}(3\cdot 5^5t_0^4t_6 - 5^5t_0^3t_5 - 2t_2t_5 + 3t_3t_6) \end{cases} \tag{2.9}$$

and

$$\mathsf{Y} = \frac{5^8(t_4 - t_0^5)^2}{t_5^3} \tag{2.10}$$

is the Yukawa coupling.

For a proof and details see §3. In the context of special geometry in String Theory, we have the choice of a coordinate system q in a local patch of the classical moduli space of mirror quintics and the differential forms α_i's such that $\nabla_{q\frac{\partial}{\partial q}}$ acts on α_i as in (2.7), see [CdlO91, CdlOGP91, Str90, CDF+97, Ali13] for the Physics literature on this and [Del97] and [CK99] page 108 for a more Hodge theoretic description of such coordinate systems for arbitrary Calabi-Yau threefolds. We would like to emphasize that Theorem 2 does not follow from period manipulations of special geometry. First of all, in our context we have the problem of the construction of the moduli space T, which is completely absent in the literature. Second, the interpretation of the vector field $q\frac{\partial}{\partial q}$ in a local patch of the classical moduli of mirror quintics, as a global algebraic vector field in the larger moduli space T, is not a triviality. More importantly notice that the vector field R_0 is defined over $\mathbb{Q}$ and this cannot be proved using transcendental methods of special geometry. The relation between our approach with those in the mentioned references are explained in [AMSY14] for arbitrary non-rigid compact Calabi-Yau threefolds.

It is convenient to consider the following weights for t_i:

$$\deg(t_i) := 3(i+1),\ i = 0,1,2,3,4,\ \deg(t_5) := 11,\ \deg(t_6) := 8. \tag{2.11}$$

In this way the right hand side of R_0 we have homogeneous rational functions of degree 4, 7, 10, 13, 16, 12, 9 which is compatible with the left hand side if we assume that the derivation increases the degree by one. The ordinary differential equation R_0 is a generalization of the Ramanujan differential equation between Eisenstein series, see for instance [Zag08, Mov12b]. Don Zagier pointed out that using the parameters

$$\begin{aligned} t_7 &:= \frac{(5^5t_0^4 + t_3)t_6 - (5^5t_0^3 + t_2)t_5}{5^4(t_4 - t_0^5)}, \\ t_8 &:= \frac{5^4(t_0^5 - t_4)}{t_5}, \\ t_9 &:= \frac{-5^5t_0^4 - t_3}{t_5}, \end{aligned} \tag{2.12}$$

the differential equation (2.9) must look simpler. I was able to rewrite it in the following way:

$$\begin{cases} \dot{t}_0 = t_8 - t_0 t_9, \\ \dot{t}_1 = -t_1 t_9 - 5^4 t_0 t_8, \\ \dot{t}_2 = -t_1 t_8 - 2t_2 t_9 - 3\cdot 5^5 t_0^2 t_8, \\ \dot{t}_3 = 4t_2 t_8 - 3t_3 t_9 - 5(t_7 t_8 - t_9 t_6) t_8, \\ \dot{t}_4 = -5t_4 t_9, \\ \dot{t}_5 = -t_6 t_8 - 3t_5 t_9 - t_3, \\ \dot{t}_6 = -2t_6 t_9 - t_2 - t_7 t_8, \\ \dot{t}_7 = -t_7 t_9 - t_1, \\ \dot{t}_8 = \frac{t_8^2}{t_5} t_6 - 3t_8 t_9, \\ \dot{t}_9 = \frac{t_8^2}{t_5} t_7 - t_9^2. \end{cases} \tag{2.13}$$

The algebra of (algebraic) differential Calabi-Yau modular forms (attached to mirror quintic) is by definition

$$\mathscr{O}_{\mathsf{T}} := \mathsf{k}[t_0, t_1, \ldots, t_6, \frac{1}{t_5(t_4 - t_0^5)t_4}]. \tag{2.14}$$

We find out that $\mathscr{O}_{\mathsf{T}}$ is a differential algebra with two different differential operators R_0 and $\check{\mathsf{R}}_0$, where the R_0 is given in (2.9) and

$$\check{\mathsf{R}}_0 := \mathsf{R}_0 + \frac{625(t_4 - t_0^5)}{t_5} \frac{\partial}{\partial t_6}. \tag{2.15}$$

Theorem 3 *We have*

$$\nabla_{\check{\mathsf{R}}_0} \begin{pmatrix} \alpha_1 \\ \alpha_2 \\ \alpha_3 \\ \alpha_4 \end{pmatrix} = \begin{pmatrix} 0 & 1 & 0 & 0 \\ 0 & 0 & \frac{5^8 (t_4 - t_0^5)^2}{t_5^3} & 0 \\ 0 & -1 & 0 & -1 \\ 0 & 0 & 0 & 0 \end{pmatrix} \begin{pmatrix} \alpha_1 \\ \alpha_2 \\ \alpha_3 \\ \alpha_4 \end{pmatrix} \tag{2.16}$$

and $\check{\mathsf{R}}_0$ is uniquely characterized by this property.

The proof is purely computational. For more details see §3.11. We sometimes call R_0 and $\check{\mathsf{R}}_0$ modular differential equations, see [Mov11b] for further examples. The reason why we need the vector field $\check{\mathsf{R}}_0$ is the following. The vector field R_0 is the algebraic incarnation of derivation of period quantities with respect to any mirror map $\frac{a}{b}$ with $\langle a, b \rangle = 0$. The vector field $\check{\mathsf{R}}_0$ does the same job when $\langle a, b \rangle = 1$. For further details see §4.15.

2.4 Geometric differential Calabi-Yau modular forms

It is a classical fact that a modular form f, and in particular the Eisenstein series E_i, of weight $i \geq 4$, can be interpreted as a function from the set of pairs (X, α) to

k, where k is a field of characteristic zero, X is an elliptic curve over k and α is a regular differential 1-form on X. Further, it satisfies the functional equation

$$f(X,k\alpha) = k^{-i}f(X,\alpha),\ k \in \mathsf{k}.$$

However, it is not so well-known that the Eisenstein series E_2 can be interpreted as a function from the set of pairs $(X,[\alpha,\omega])$ to k such that

$$f(X,[k\alpha,k^{-1}\omega]) = k^{-2}f(X,[\alpha,\omega]),\ \ k \in \mathsf{k}-\{0\}, \tag{2.17}$$
$$f(X,[\alpha,\omega+k'\alpha]) = k^{-2}f(X,[\alpha,\omega])+k'k^{-1},\ \ k' \in \mathsf{k}-\{0\}, \tag{2.18}$$

see [Mov12b]. Here, X is as before and α,ω form a basis of the de Rham cohomology $H^1_{\mathrm{dR}}(X)$ such that α is represented by a regular differential form and $\langle\alpha,\omega\rangle = 1$. In general modular forms and quasi-modular forms in the sense of Kaneko-Zagier [KZ95] or differential modular forms in the sense of [Mov08a], turn out to be regular functions in the moduli T of enhanced elliptic curves. Therefore, the $\mathbb{Q}$-algebra $\mathscr{O}_{\mathsf{T}}$ of regular functions in T, turns out to be the algebra of quasi-modular forms.

We start our definition of differential Calabi-Yau modular forms by generalizing this geometric approach. Let T be the moduli of enhanced mirror quintics. A differential Calabi-Yau modular form f is just a regular function on T, that is, $f \in \mathscr{O}_{\mathsf{T}}$. There is an algebraic group G of dimension 6 which acts on T and hence on $\mathscr{O}_{\mathsf{T}}$, see §3.10. Using this algebraic group, one can define many gradings of $\mathscr{O}_{\mathsf{T}}$. In the case of elliptic curves G is of dimension two and it has one multiplicative subgroup and one additive subgroup, however, in the mirror quintic case G has two multiplicative groups and 4 additive groups.

Let us state some sample of functional equations in our context. The regular functions $t_i,\ \ i=0,4$, respectively the Yukawa coupling Y, can be interpreted as functions from the set of pairs (X,α_1), respectively $(X,[\alpha_1,\alpha_2])$, to k such that

$$t_i(X,k_1\alpha_1) = k_1^{-(i+1)}t_i(X,\alpha_1),\ \ \ i=0,4,$$

$$\mathsf{Y}(X,[k_1\alpha_1,k_2\alpha_2]) = k_1^{-1}k_2^3\mathsf{Y}(X,[\alpha_1,\alpha_2]),\ \ k_1,k_2 \in \mathsf{k}-\{0\}.$$

Here, X is a mirror quintic Calabi-Yau threefold and $\alpha_1 \in F^3,\ \alpha_2 \in F^2$, where $F^i \subset H^3_{\mathrm{dR}}(X)$ are the pieces of the Hodge filtration in (2.1). A similar statement is also valid for all t_i's, see §3.10.

2.5 Eisenstein series

In the case of elliptic curves the (generalized) period domain is defined to be

$$\mathsf{U} := \mathrm{SL}(2,\mathbb{Z})\backslash\Pi,$$

where

$$\Pi := \left\{ \begin{pmatrix} x_1 & x_2 \\ x_3 & x_4 \end{pmatrix} \in \mathrm{SL}(2,\mathbb{C}) \mid \mathrm{Im}(x_1\overline{x_3}) > 0 \right\}.$$

The period map

$$\mathsf{P} : \mathsf{T} \to \mathsf{U}, \tag{2.19}$$

$$t \mapsto \left[\frac{1}{\sqrt{-2\pi i}} \begin{pmatrix} \int_{\delta_1} \frac{dx}{y} & \int_{\delta_1} \frac{xdx}{y} \\ \int_{\delta_2} \frac{dx}{y} & \int_{\delta_2} \frac{xdx}{y} \end{pmatrix} \right]$$

turns out to be a biholomorphism, however, for generalizations of modular forms we do not need this fact. Here, $[\cdot]$ means the equivalence class and $\{\delta_1, \delta_2\}$ is a basis of the $\mathbb{Z}$-module $H_1(E_t, \mathbb{Z})$ with $\langle \delta_1, \delta_2 \rangle = -1$. We consider the composition of maps

$$\mathbb{H} \overset{i}{\hookrightarrow} \Pi \to \mathsf{U} \overset{\mathsf{P}^{-1}}{\to} \mathsf{T}(\mathbb{C}) \tag{2.20}$$

where $\mathbb{H} = \{\tau \in \mathbb{C} \mid \mathrm{Im}(\tau) > 0\}$ is the upper half-plane,

$$i : \mathbb{H} \to \Pi,\ i(\tau) = \begin{pmatrix} \tau & -1 \\ 1 & 0 \end{pmatrix},$$

and $\Pi \to \mathsf{U}$ is the quotient map. The pullback of the regular function $t_i,\ \ i = 1,2,3$ in $\mathsf{T}(\mathbb{C})$ by the composition map $\mathbb{H} \to \mathsf{T}(\mathbb{C})$ is the Eisenstein series

$$E_{2i} := a_i \left(1 - \frac{4i}{B_{2i}} \sum_{n=1}^{\infty} \left(\sum_{d|n} d^{2i-1} \right) q^n \right), \tag{2.21}$$

where $B_2 = \frac{1}{6}$, $B_4 = -\frac{1}{30}$, $B_6 = \frac{1}{42}$, $\ldots$ are Bernoulli numbers and

$$(a_1, a_2, a_3) = \left(\frac{2\pi i}{12}, 12\left(\frac{2\pi i}{12}\right)^2, 8\left(\frac{2\pi i}{12}\right)^3\right).$$

It is not at all clear why the above procedure leads us to the Eisenstein series. However, we can modify the above procedure into another one which computes the coefficients of the q-expansions of E_i's without explaining why the arithmetic sum $\sum_{d|n} d^{i-1}$ arises. This is explained in the next section. We will use the same idea in the case of mirror quintics. In §4 we define the period domain and the period map in the case of mirror quintics. Despite the fact that the period map is neither surjective nor injective, we generalize the above argument to the case of mirror quintics. The inverse of the period map P is considered to be a local map defined in some region in the image of P. In this way to each t_i we attach a holomorphic function $t_i(\tau_0)$, where τ_0 is defined in some neighborhood of $\sqrt{-1}\infty$. In order to do q-expansions we take one parameter family of mirror quintic Calabi-Yau threefolds, compute its periods and Gauss-Manin connection/Picard-Fuchs equation and express t_i's in terms of periods. The outcome of all these computations is presented in §2.7.

2.6 Elliptic integrals and modular forms

Since the moduli space of elliptic curves is one-dimensional, we take a one parameter family of elliptic curves, compute its periods and recover the Eisenstein series in §2.5. For instance, we take

$$E_z : y^2 - 4x^3 + 12x - 8(2z-1) = 0$$

and find out that

$$\begin{pmatrix} \int_{\delta_1} \frac{dx}{y} & \int_{\delta_2} \frac{dx}{y} \\ \int_{\delta_1} \frac{xdx}{y} & \int_{\delta_2} \frac{xdx}{y} \end{pmatrix} = \begin{pmatrix} \frac{\pi i}{\sqrt{3}} F(\frac{1}{6},\frac{5}{6},1|1-z) & \frac{\pi}{\sqrt{3}} F(\frac{1}{6},\frac{5}{6},1|z) \\ \frac{\pi i}{\sqrt{3}} F(\frac{-1}{6},\frac{7}{6},1|1-z) & -\frac{\pi}{\sqrt{3}} F(\frac{-1}{6},\frac{7}{6},1|z) \end{pmatrix},$$

where

$$F(a,b,c|z) = \sum_{n=0}^{\infty} \frac{(a)_n (b)_n}{(c)_n n!} z^n, \; c \notin \{0,-1,-2,-3,\dots\} \tag{2.22}$$

is the Gauss hypergeometric function, $(a)_n := a(a+1)(a+2)\cdots(a+n-1)$ and $\delta_1, \delta_2 \in H_1(E_\psi, \mathbb{Z})$ are two cycles which can be described precisely. In practice, one only needs the differential equation satisfied by the elliptic integrals $\int_\delta \frac{dx}{y}$, $\delta \in H_1(E_z, \mathbb{Z})$. This is the Picard-Fuchs equation

$$\frac{5}{36} I + 2\psi I' + (\psi^2 - 4) I'' = 0. \tag{2.23}$$

The discussion in §2.5 gives us the following equalities:

$$F\left(-\frac{1}{6},\frac{7}{6},1 \mid z\right) F\left(\frac{1}{6},\frac{5}{6},1 \mid z\right) = E_2\left(i\frac{F\left(\frac{1}{6},\frac{5}{6},1|1-z\right)}{F\left(\frac{1}{6},\frac{5}{6},1|z\right)}\right), \tag{2.24}$$

$$F\left(\frac{1}{6},\frac{5}{6},1 \mid z\right)^4 = E_4\left(i\frac{F(\frac{1}{6},\frac{5}{6},1|1-z)}{F(\frac{1}{6},\frac{5}{6},1|z)}\right),$$

$$(1-2z)F\left(\frac{1}{6},\frac{5}{6},1 \mid z\right)^6 = E_6\left(i\frac{F(\frac{1}{6},\frac{5}{6},1|1-z)}{F(\frac{1}{6},\frac{5}{6},1|z)}\right).$$

The lesson from these equalities is simple. If in the left hand side of the above equalities we replace z with the inverse of the Schwarz map (or mirror map in Physics literature)

$$\tau := i\frac{F(\frac{1}{6},\frac{5}{6},1|1-z)}{F(\frac{1}{6},\frac{5}{6},1|z)}$$

then we get the well-known Eisenstein series. Actually the above type equalities are quite old. For instance, the following equality is due to Fricke and Klein:

$$\sqrt[4]{E_4(\tau)} = F(\frac{1}{12}, \frac{5}{12}, 1; \frac{1728}{j(\tau)}),$$

where $j = 1728\frac{E_4^3}{E_4^3 - E_6^2} = q^{-1} + 744 + 196884q + \cdots$ is the j-function. For further equalities of this type see [LY96b], I. Hodge theory of elliptic curves implies that $\mathrm{Im}(\tau) > 0$, and actually, the image of τ covers the upper half plane, see §10.1. Similar inequalities in the case of mirror quintics are computed in §4.3.

2.7 Periods and differential Calabi-Yau modular forms, I

We would like to repeat the same line of reasoning used in §2.5 and §2.6 and produce q-expansions for mirror quintics. It turns out that we only need the Picard-Fuchs equation of the periods of the holomorphic $(3,0)$ form η of a one parameter family X_z of mirror quintics, see §3.1. This is the following fourth-order linear differential equation:

$$\theta^4 - z(\theta + \frac{1}{5})(\theta + \frac{2}{5})(\theta + \frac{3}{5})(\theta + \frac{4}{5}) = 0, \quad \theta = z\frac{\partial}{\partial z}. \tag{2.25}$$

That is the left hand side of (2.25) annihilates the periods of η. This differential equation was first calculated in [CdlOGP91]. A basis of the solution space of (2.25) is given by

$$\psi_i(z) = \frac{1}{i!}\frac{\partial^i}{\partial \varepsilon^i}(5^{-5\varepsilon}F(\varepsilon, z)), \quad i = 0, 1, 2, 3,$$

where

$$F(\varepsilon, z) := \sum_{n=0}^{\infty} \frac{(\frac{1}{5}+\varepsilon)_n(\frac{2}{5}+\varepsilon)_n(\frac{3}{5}+\varepsilon)_n(\frac{4}{5}+\varepsilon)_n}{(1+\varepsilon)_n^4} z^{\varepsilon+n}$$

and $(a)_n := a(a+1)\cdots(a+n-1)$ for $n > 0$ and $(a)_0 := 1$. We use the base change

$$\begin{pmatrix} x_{11} \\ x_{21} \\ x_{31} \\ x_{41} \end{pmatrix} = \begin{pmatrix} 0 & 0 & 1 & 0 \\ 0 & 0 & 0 & 1 \\ 0 & 5 & \frac{5}{2} & -\frac{25}{12} \\ -5 & 0 & -\frac{25}{12} & 200\frac{\zeta(3)}{(2\pi i)^3} \end{pmatrix} \begin{pmatrix} \frac{1}{5^4}\psi_3 \\ \frac{2\pi i}{5^4}\psi_2 \\ \frac{(2\pi i)^2}{5^4}\psi_1 \\ \frac{(2\pi i)^3}{5^4}\psi_0 \end{pmatrix}.$$

We have $x_{i1} = \int_{\delta_i} \eta, \ i = 1,2,3,4$, where η is a holomorphic three form on X_z and $\delta_i \in H_3(X_z, \mathbb{Z}), \ i = 1,2,3,4$ is a symplectic basis, see §3.1 and §4.10. That is why we do such a base change. It turns out that in the new basis x_{1i} the monodromy of (2.25) around the singularities $z = 0$ and $z = 1$ are respectively given by

$$M_0 := \begin{pmatrix} 1 & 1 & 0 & 0 \\ 0 & 1 & 0 & 0 \\ 5 & 5 & 1 & 0 \\ 0 & -5 & -1 & 1 \end{pmatrix}, \quad M_1 := \begin{pmatrix} 1 & 0 & 0 & 0 \\ 0 & 1 & 0 & 1 \\ 0 & 0 & 1 & 0 \\ 0 & 0 & 0 & 1 \end{pmatrix}, \tag{2.26}$$

that is, the analytic continuation of the 4×1 matrix $[x_{i1}]$ around the singularity $z=0$ (respectively $z=1$) is given by $M_0[x_{i1}]$, respectively $M_1[x_{i1}]$ (see for instance [DM06], [vEvS06] and [CYY08] for similar calculations).

Let

$$\tau_0 := \frac{x_{11}}{x_{21}}, \quad q := e^{2\pi i \tau_0},$$

and

$$x_{ij} := \theta^{j-1} x_{i1}, \quad i,j = 1,2,3,4.$$

The multi-valued map τ_0 is called the mirror map in the Physics literature. The holomorphic function $q : (\mathbb{C},0) \to (\mathbb{C},0)$ turns out to be invertible and so it can be served as a coordinate system in $(\mathbb{C},0)$. Therefore, any holomorphic function in $(\mathbb{C},0)$ can be written as a function of q. In the theorem below we introduce seven such functions.

Theorem 4 *The seven functions*

$$\begin{aligned}
t_0 &= x_{21}, \qquad (2.27)\\
t_1 &= 5^4 x_{21}\left((6z-1)x_{21} + 5(11z-1)x_{22} + 25(6z-1)x_{23} + 125(z-1)x_{24}\right),\\
t_2 &= 5^4 x_{21}^2\left((2z-7)x_{21} + 15(z-1)x_{22} + 25(z-1)x_{23}\right),\\
t_3 &= 5^4 x_{21}^3\left((z-6)x_{21} + 5(z-1)x_{22}\right),\\
t_4 &= z x_{21}^5,\\
t_5 &= 5^5(z-1)x_{21}^2\left(x_{12}x_{21} - x_{11}x_{22}\right),\\
t_6 &= 5^5(z-1)x_{21}\left(3(x_{12}x_{21} - x_{11}x_{22}) + 5(x_{13}x_{21} - x_{11}x_{23})\right)
\end{aligned}$$

are holomorphic at $z=0$ and so there are holomorphic functions h_i defined in some neighborhood of $0 \in \mathbb{C}$ such that

$$t_i = \left(\frac{2\pi i}{5}\right)^{d_i} h_i(e^{2\pi i \tau_0}), \qquad (2.28)$$

where

$$d_i := 3(i+1),\ i = 0,1,2,3,4,\ \ d_5 := 11,\ \ d_6 := 8.$$

Moreover, t_i's satisfy the ordinary differential equation (2.9), with $\dot{*} := \frac{\partial *}{\partial \tau_0}$.

The proof of the above theorem as it is, is just a mere computation. The reader may wonder why we take such a strange combination of periods in t_i. The answer is that they are obtained after a geometric engineering in §4. From now on we write t_i's for either regular functions on T, period expressions (2.27) or the q-expansions h_i; being clear in the text which we mean.

The ordinary differential equation (2.9) is a natural generalization of the Darboux-Halphen-Ramanujan differential equations (see [Mov11b] for more details on these differential equations). Halphen in [Hal81] and subsequent papers already noticed that if one knows at least one solution of such vector fields then one can construct all other solutions using an action of $\mathrm{SL}(2,\mathbb{C})$. This is known as the Halphen property. In §4.15 we generalize this property for (2.9) replacing $\mathrm{SL}(2,\mathbb{C})$ with $\mathrm{Sp}(4,\mathbb{C})$. In this way, various q-expansions attached to $f \in \mathbb{C}[t]$ are related to each other through such an action.

2.8 Integrality of Fourier coefficients

In Theorem 4, we write each t_i as a formal power series in q, $t_i = \sum_{n=0}^{\infty} t_{i,n} q^n$ and substitute in (2.9) with $* = 5q\frac{\partial *}{\partial q}$ and we see that it determines all the coefficients $t_{i,n}$ uniquely with the initial values:

$$t_{0,0} = \frac{1}{5},\ t_{0,1} = 24,\ t_{4,0} = 0 \tag{2.29}$$

and assuming that $t_{5,0} \neq 0$. In fact the differential equation (2.9) seems to be the simplest way of writing the mixed recursion between $t_{i,n}$, $n \geq 2$. Some of the first coefficients of t_i's are given in the table below:

	q^0	q^1	q^2	q^3	q^4	q^5	q^6
$\frac{1}{24}t_0$	$\frac{1}{120}$	1	175	117625	111784375	126958105626	160715581780591
$\frac{-1}{750}t_1$	$\frac{1}{30}$	3	930	566375	526770000	592132503858	745012928951258
$\frac{-1}{50}t_2$	$\frac{7}{10}$	107	50390	29007975	26014527500	28743493632402	35790559257796542
$\frac{-1}{5}t_3$	$\frac{6}{5}$	71	188330	100324275	86097977000	93009679497426	114266677893238146
$-t_4$	0	-1	170	41475	32183000	32678171250	38612049889554
$\frac{1}{125}t_5$	$-\frac{1}{125}$	15	938	587805	525369650	577718296190	716515428667010
$\frac{1}{25}t_6$	$-\frac{3}{5}$	187	28760	16677425	15028305250	16597280453022	20644227272244012
$\frac{1}{125}t_7$	$-\frac{1}{5}$	13	2860	1855775	1750773750	1981335668498	2502724752660128
$\frac{1}{10}t_8$	$-\frac{1}{50}$	13	6425	6744325	8719953625	12525150549888	19171976431076873
$\frac{1}{10}t_9$	$-\frac{1}{10}$	17	11185	12261425	16166719625	23478405649152	36191848368238417

The integrality of the coefficients of t_i's are not at clear from the recursion given by R_0 or by the period manipulation in Theorem 4. Both methods imply at most that such coefficients are rational numbers.

Theorem 5 *All the coefficients of q^i, $i \geq 1$ in the formal power series*

$$\frac{1}{24}t_0,\ \frac{-1}{750}t_1,\ \frac{-1}{50}t_3,\ \frac{-1}{5}t_3,\ -t_4,\ \frac{1}{125}t_5,\ \frac{1}{25}t_6,\ \frac{1}{125}t_7,\ \frac{1}{10}t_8,\ \frac{1}{10}t_9$$

are integers.

This kind of integrality properties was first proved in many particular cases by Lian-Yau [LY96a, LY96b] and conjectured for hypersurfaces in toric projective spaces by Hosono-Lian-Yau in [HLY96]. The general format in the framework of hypergeometric functions was formulated by Zudilin in [Zud02] and finally established by Krattenthaler-Rivoal in [KRNT]. All these methods are based on a theorem of Dwork. For a proof of Theorem 5 and further discussion on the integrality of the coefficients of q-expansions see Appendix C. The Yukawa coupling has the following q-expansion:

$$\mathsf{Y} = (\frac{2\pi i}{5})^{-3}\left(5+2875\frac{q}{1-q}+609250\cdot 2^3\frac{q^2}{1-q^2}+\cdots+n_d d^3\frac{q^d}{1-q^d}+\cdots\right). \tag{2.30}$$

Here, n_d is the virtual number of rational curves in a generic quintic threefold [Supp Item 2]. The numbers n_d are also called instanton numbers or BPS degeneracies.

In the case of elliptic curves we can prove that E_2, E_4, E_6 are algebraically independent over $\mathbb{C}$ and in the context of the present text we have

Theorem 6 *The functions t_i, $i = 0, 1, \ldots, 6$ as formal power series in q are algebraically independent over $\mathbb{C}$, this means that there is no polynomial P in seven variables and with coefficients in $\mathbb{C}$ such that $P(t_0, t_1, \ldots, t_6) = 0$.*

For a proof see §4.11. In the case of mirror quintics we have different type of q-expansions attached to t_i's. This is mainly due to different choices of mirro maps. For more details see §5 and §11.9.

2.9 Quasi- or differential modular forms

The reader who is expert in classical modular forms may ask for the functional equations of t_i's. Recall that in almost all books on modular forms we find the following definition: A holomorphic function $f : \mathbb{H} \to \mathbb{C}$ is called a modular form of weight k for the group $\mathrm{SL}(2,\mathbb{Z})$ (full modular form of weight k) if

1. f has a finite growth at infinity, i.e.

$$\lim_{\mathrm{Im}\,\tau\to\infty} f(\tau) = a_\infty \in \mathbb{C}, \tag{2.31}$$

2. f satisfies

$$(c\tau+d)^{-k} f(\frac{a\tau+b}{c\tau+d}) = f(\tau), \tag{2.32}$$

$$\forall \begin{pmatrix} a & b \\ c & d \end{pmatrix} \in \mathrm{SL}(2,\mathbb{Z}),\ \tau \in \mathbb{H}.$$

It turns out that the graded $\mathbb{C}$-algebra of full modular forms is the polynomial ring $\mathbb{C}[E_4, E_6]$ with $\mathrm{weight}(E_i) = i,\ \ i = 4, 6$. The functional equation of quasi-modular forms has more than two terms, for instance E_2 satisfies:

$$(c\tau+d)^{-2} E_2(\frac{a\tau+b}{c\tau+d}) = E_2(\tau) + c(c\tau+d)^{-1}. \tag{2.33}$$

However, it is possible to manage this and prove that the graded algebra of quasi-modular forms is the polynomial ring $\mathbb{C}[E_2, E_4, E_6]$ with $\mathrm{weight}(E_i) = i,\ \ i = 2, 4, 6$, see [KZ95, MR05]. A geometric proof of this fact is based on the simple observation that the functional equations (2.17) and (2.18) are the algebraic incarnation of (2.32) and (2.33), see for instance [Mov08a, Mov12b]. The group $\mathrm{SL}(2,\mathbb{Z})$ can be interpreted as the monodromy group of the family of elliptic curves (2.3). In order to generalize this picture for mirror quintics we have to reinterpret the upper half plane $\mathbb{H}$ in terms of elliptic integrals, see §10.1. It turns out that it is quite hard, and apparently useless, to derive the algebra $\mathscr{O}_{\mathsf{T}}$ attached to mirror quintics using functional equations. We explain this in §2.10.

What we develop in the present text under the name of differential Calabi-Yau modular forms, is actually a generalization of (elliptic) quasi-modular forms. One might be interested to know what is a possible generalization of (elliptic) modular forms in the context of Calabi-Yau varieties. In §10 we discuss the fact that t_0 and t_4 are true generalizations of the Eisenstein series E_4 and E_6 in the context of mirror quintics. The rest t_1,t_2,t_3,t_5,t_6 might be considered as different realizations of E_2. This distinction might be important in the future, as most of the interesting applications in the literature, such as arithmetic modularity theorem and monstrous moonshine, use modular forms and not quasi-modular forms. We reserve the name Calabi-Yau modular form for the elements of the polynomial ring $\mathbb{C}[t_0,t_4]$ and its generalizations in the context of Calabi-Yau threefolds as in [AMSY14], see also §3.2.

2.10 Functional equations

Let $\mathbb{H}$ be the monodromy covering of $(\mathbb{C}-\{0,1\})\cup\{\infty\}$ associated to the monodromy group $\Gamma := \langle M_1, M_0\rangle$ of (2.25) (see §4.6). The set $\mathbb{H}$ is biholomorphic to the upper half plane. This is equivalent to say that the only relation between M_0 and M_1 is $(M_0M_1)^5 = I$, see for instance [BT14]. We do not need the mentioned fact because we do not need the coordinate system on $\mathbb{H}$ given by this biholomorphism. We are going to use local coordinate systems on $\mathbb{H}$ given by the period of mirror quintics. The monodromy group Γ acts from the left on $\mathbb{H}$ in a canonical way:

$$(A,w) \mapsto A(w) \in \mathbb{H}, \quad A \in \Gamma, \ w \in \mathbb{H}$$

and the quotient $\Gamma\backslash\mathbb{H}$ is biholomorphic to $(\mathbb{C}-\{0,1\})\cup\{\infty\}$. This action has one elliptic point ∞ of order 5 and two cusps 0 and 1. We can regard x_{ij} as holomorphic one valued functions on $\mathbb{H}$. For simplicity we use the same notation for these functions: $x_{ij} : \mathbb{H} \to \mathbb{C}$. We define

$$\tau_i : \mathbb{H} \to \mathbb{C}, \quad \tau_0 := \frac{x_{11}}{x_{21}}, \ \tau_1 := \frac{x_{31}}{x_{21}}, \ \tau_2 := \frac{x_{41}}{x_{21}}, \ \tau_3 := \frac{x_{31}x_{22} - x_{32}x_{21}}{x_{11}x_{22} - x_{12}x_{21}},$$

which are a priori meromorphic functions on $\mathbb{H}$. We will use τ_0 as a local coordinate around a point $w \in \mathbb{H}$ whenever w is not a pole of τ_0 and the derivative of τ_0 does not vanish at w. In this way we need to express τ_i, $i = 1,2,3$ as functions of τ_0:

$$\tau_1 = -\frac{25}{12} + \frac{5}{2}\tau_0(\tau_0+1) + \frac{\partial H}{\partial \tau_0},$$
$$\tau_2 = 200\frac{\zeta(3)}{(2\pi i)^3} - \frac{5}{6}\tau_0(\frac{5}{2}+\tau_0^2) - \tau_0\frac{\partial H}{\partial \tau_0} - 2H,$$
$$\tau_3 = \frac{\partial \tau_1}{\partial \tau_0}, \tag{2.34}$$

where

$$H = \frac{1}{(2\pi i)^3}\sum_{n=1}^{\infty}(\sum_{d|n} n_d d^3)\frac{e^{2\pi i \tau_0 n}}{n^3} \tag{2.35}$$

and n_d is the virtual number of rational curves of degree d in a generic quintic threefold. A complete description of the image of τ_0 is not yet known. Now, t_i's are well-defined holomorphic functions on $\mathbb{H}$. The functional equations of t_i's with respect to the action of an arbitrary element of Γ are complicated mixed equalities which we have described in §4.6. Since Γ is generated by M_0 and M_1 it is enough to explain them for these two elements. The functional equations of t_i's with respect to the action of M_0 and written in the τ_0-coordinate are the trivial equalities $t_i(\tau_0) = t_i(\tau_0+1)$, $i = 0,1,\ldots,6$.

Theorem 7 *With respect to the action of M_1, t_i's written in the τ_0-coordinate satisfy the following functional equations:*

$$t_0(\tau_0) = t_0(\frac{\tau_0}{\tau_2+1})\frac{1}{\tau_2+1},$$
$$t_1(\tau_0) = t_1(\frac{\tau_0}{\tau_2+1})\frac{1}{(\tau_2+1)^2} + t_7(\frac{\tau_0}{\tau_2+1})\frac{\tau_0\tau_3-\tau_1}{(\tau_2+1)(\tau_0^2\tau_3-\tau_0\tau_1+\tau_2+1)}$$
$$+t_9(\frac{\tau_0}{\tau_2+1})\frac{-\tau_0}{(\tau_2+1)^2} + \frac{1}{\tau_2+1},$$
$$t_2(\tau_0) = t_2(\frac{\tau_0}{\tau_2+1})\frac{1}{(\tau_2+1)^3} + t_6(\frac{\tau_0}{\tau_2+1})\frac{\tau_0\tau_3-\tau_1}{(\tau_2+1)^2(\tau_0^2\tau_3-\tau_0\tau_1+\tau_2+1)}$$
$$+t_8(\frac{\tau_0}{\tau_2+1})\frac{-\tau_0}{(\tau_2+1)^3},$$
$$t_3(\tau_0) = t_3(\frac{\tau_0}{\tau_2+1})\frac{1}{(\tau_2+1)^4} + t_5(\frac{\tau_0}{\tau_2+1})\frac{\tau_0\tau_3-\tau_1}{(\tau_2+1)^3(\tau_0^2\tau_3-\tau_0\tau_1+\tau_2+1)},$$
$$t_4(\tau_0) = t_4(\frac{\tau_0}{\tau_2+1})\frac{1}{(\tau_2+1)^5},$$
$$t_5(\tau_0) = t_5(\frac{\tau_0}{\tau_2+1})\frac{1}{(\tau_2+1)^2(\tau_0^2\tau_3-\tau_0\tau_1+\tau_2+1)},$$
$$t_6(\tau_0) = t_6(\frac{\tau_0}{\tau_2+1})\frac{1}{(\tau_2+1)(\tau_0^2\tau_3-\tau_0\tau_1+\tau_2+1)} + t_8(\frac{\tau_0}{\tau_2+1})\frac{\tau_0^2}{(\tau_2+1)^3},$$

where t_7, t_8, t_9 are given in (2.12).

For a proof see §4.8. Since we do not know the global behavior of τ_0, the above equalities must be interpreted in the following way: for any fixed branch of $t_i(\tau_0)$ there is a path γ in the image of $\tau_0 : \mathbb{H} \to \mathbb{C}$ which connects τ_0 to $\frac{\tau_0}{\tau_2+1}$ and such that the analytic continuation of t_i's along the path γ satisfy the above equalities. For this reason it may be reasonable to use a new name for all t_i's in the right hand side of the equalities in Theorem 2. We could also state Theorem 2 without using any local coordinate system on $\mathbb{H}$: we regard t_i as holomorphic functions on $\mathbb{H}$ and, for instance, the first equality in Theorem 2 can be derived from the equalities:

$$t_0(w) = t_0(M_1(w))\frac{1}{\tau_2(w)+1}, \quad \tau_0(M_1(w)) = \frac{\tau_0(w)}{\tau_2(w)+1}.$$

The Yukawa coupling Y defined in (2.10) turns out to satisfy the functional equation

$$\mathsf{Y}(\tau_0) = \mathsf{Y}(\frac{\tau_0}{\tau_2+1})\frac{(\tau_0^2\tau_3 - \tau_0\tau_1 + \tau_2 + 1)^3}{(\tau_2+1)^4}.$$

The functional equations in Theorem 2 are direct translations of the algebraic functional equations described in §2.4, and so, they must be considered as new modularity equations satisfied by holomorphic differential Calabi-Yau modular forms. All the attempts in order to find classical modularity for the Yukawa coupling (the only differential Calabi-Yau modular form studied in the literature) has not been satisfactory. For instance, in [ABKv2, Zud11] we find a kind of $\mathrm{Sp}(4,\mathbb{Z})$-modularity with either functions not defined in the Siegel upper half plane or defined through non-holomorphic embeddings.

2.11 Conifold singularity

The space of classical (or elliptic) modular or quasi-modular forms of a fixed degree for discrete subgroups of $\mathrm{SL}(2,\mathbb{R})$, is a finite dimensional vector space. This simple observation is the origin of many number theoretic applications (see for instance [Zag08]) and we may try to generalize such applications to the context of present text. Having this in mind, we have to describe the behavior of t_i's around the conifold cusp $z = 1$ of the Picard-Fuchs equation (2.25).

So far, we have described 7 holomorphic quantities $t_i, i = 0,1,2,\ldots,6$, either as a function of z or q, and we have explained how they can be considered as generalization of classical Eisenstein series E_{2i}, $i = 1,2,3$. All these quantities are studied in a neighborhood of the maximal unipotent cusp $z = 0$ (MUM for short), where we have basically one choice for the mirror map (a generalization of Schwarz map in our context). Around the conifold cusp and to each $f \in \mathscr{O}_\mathsf{T}$ we can associate different q-expansions which correspond to different choices of the mirror map. For all these q-expansions we do not have a closed formula, but a recursion given by R_0 and $\check{\mathsf{R}}_0$, and some initial data calculated using the asymptotic behavior of periods. Around the maximal unipotent cusp we get q-expansion with integer coefficients. From this data we can compute Gromov-Witten and instanton numbers. However, around the conifold cusp we get rational coefficients (with infinitely many primes appearing in their denominators) and no enumerative properties for these q-expansions are known. For more details see §4.18, §4.18.

2.12 The Lie algebra $\mathfrak{sl}_2$

Recall our notations in the case of elliptic curves in §2.2. The $\mathbb{C}$-vector space generated by the vector fields

$$e = -(t_1^2 - \frac{1}{12}t_2)\frac{\partial}{\partial t_1} - (4t_1t_2 - 6t_3)\frac{\partial}{\partial t_2} - (6t_1t_3 - \frac{1}{3}t_2^2)\frac{\partial}{\partial t_3}, \tag{2.36}$$

$$h = -6t_3\frac{\partial}{\partial t_3} - 4t_2\frac{\partial}{\partial t_2} - 2t_1\frac{\partial}{\partial t_1}, \; f = \frac{\partial}{\partial t_1}$$

and equipped with the classical bracket of vector fields is isomorphic to the Lie Algebra $\mathfrak{sl}_2$:

$$[h,e] = 2e, \quad [h,f] = -2f, \quad [e,f] = h.$$

Note that $\mathsf{R} = -e$. Once the Gauss-Manin connection matrix A in (2.5) is computed, one can verify the above statement immediately. The vector fields e,h,f are uniquely characterized by

$$\nabla_v \begin{pmatrix} \alpha \\ \omega \end{pmatrix} = \mathsf{A}_v \cdot \begin{pmatrix} \alpha \\ \omega \end{pmatrix},$$

for $v = e,h,f$, where

$$\mathsf{A}_h = \begin{pmatrix} 1 & 0 \\ 0 & -1 \end{pmatrix}, \quad \mathsf{A}_e = \begin{pmatrix} 0 & 1 \\ 0 & 0 \end{pmatrix}, \quad \mathsf{A}_f = \begin{pmatrix} 0 & 0 \\ 1 & 0 \end{pmatrix}.$$

Here, A_v is the pairing of the entries of A, which are differential 1-forms, with the vector field v. A similar discussion is also valid for the Darboux and Halphen differential equations, see for instance [Mov12a, Gui07]. The generalization of this Lie algebra for Calabi-Yau threefolds in [AMSY14] plays an important role in the algebraic version of the anomaly equation discovered in [BCOV94]. Further generalizations in the context of tt^* geometries can be found in [Ali14]. We are going to explain this in the case of mirror quintics.

2.13 BCOV holomorphic anomaly equation, I

The moduli T of varieties of a fixed topological type and enhanced with differential forms gives us a natural framework for dealing with automorphic forms and topological string partition functions in the same time. Explicit construction of such moduli spaces in the case of elliptic curves is done in [Mov12b]. For mirror quintics it is done in §3. In the case of elliptic curves we recover the theory of modular and quasi-modular forms and in the case of mirror quintics, topological string partition functions studied in [BCOV93, BCOV94] turn out to be regular functions on T. Our approach can be considered as a geometric framework for the results of Yamaguchi-Yau on polynomial structure of topological string partition functions, see [YY04].

Theorem 8 *There are unique vector fields* R_i, $i = 0, 1, 2 \ldots, 6$ *in* T *and a unique regular function* Y *on* T *such that* $\nabla_{\mathsf{R}_i}\alpha = \mathsf{A}_{\mathsf{R}_i}\alpha$, *where*

$$\mathsf{A}_{\mathsf{R}_0} = \begin{pmatrix} 0 & 1 & 0 & 0 \\ 0 & 0 & \mathsf{Y} & 0 \\ 0 & 0 & 0 & -1 \\ 0 & 0 & 0 & 0 \end{pmatrix}, \quad \mathsf{A}_{\mathsf{R}_1} = \begin{pmatrix} 0 & 0 & 0 & 0 \\ 0 & -1 & 0 & 0 \\ 0 & 0 & 1 & 0 \\ 0 & 0 & 0 & 0 \end{pmatrix}, \quad \mathsf{A}_{\mathsf{R}_2} = \begin{pmatrix} -1 & 0 & 0 & 0 \\ 0 & 0 & 0 & 0 \\ 0 & 0 & 0 & 0 \\ 0 & 0 & 0 & 1 \end{pmatrix},$$

$$\mathsf{A}_{\mathsf{R}_3} = \begin{pmatrix} 0 & 0 & 0 & 0 \\ -1 & 0 & 0 & 0 \\ 0 & 0 & 0 & 0 \\ 0 & 0 & 1 & 0 \end{pmatrix}, \quad \mathsf{A}_{\mathsf{R}_4} = \begin{pmatrix} 0 & 0 & 0 & 0 \\ 0 & 0 & 0 & 0 \\ 0 & 1 & 0 & 0 \\ 0 & 0 & 0 & 0 \end{pmatrix},$$

$$\mathsf{A}_{\mathsf{R}_5} = \begin{pmatrix} 0 & 0 & 0 & 0 \\ 0 & 0 & 0 & 0 \\ 1 & 0 & 0 & 0 \\ 0 & 1 & 0 & 0 \end{pmatrix}, \quad \mathsf{A}_{\mathsf{R}_6} = \begin{pmatrix} 0 & 0 & 0 & 0 \\ 0 & 0 & 0 & 0 \\ 0 & 0 & 0 & 0 \\ 1 & 0 & 0 & 0 \end{pmatrix}.$$

Theorem 8 for R_0 is the same as Theorem 2. In the general context of Calabi-Yau threefolds it was proved in [AMSY14]. The methods used in this article are based upon the special geometry of String Theory, whereas the proof of Theorem 8 is purely computational and hence elementary, see §6.2. The Lie bracket structure between the vector fields R_i's is given in the following table:

	R_0	R_1	R_2	R_3	R_4	R_5	R_6
R_0	0	R_0	$-\mathsf{R}_0$	$-\mathsf{R}_2+\mathsf{R}_1$	$Y\cdot\mathsf{R}_1$	$2\mathsf{R}_4+Y\cdot\mathsf{R}_3$	R_5
R_1	$-\mathsf{R}_0$	0	0	R_3	$-2\mathsf{R}_4$	$-\mathsf{R}_5$	0
R_2	R_0	0	0	$-\mathsf{R}_3$	0	$-\mathsf{R}_5$	$-2\mathsf{R}_6$
R_3	$\mathsf{R}_2-\mathsf{R}_1$	$-\mathsf{R}_3$	R_3	0	$-\mathsf{R}_5$	$-2\mathsf{R}_6$	0
R_4	$-Y\cdot\mathsf{R}_1$	$2\mathsf{R}_4$	0	R_5	0	0	0
R_5	$-2\mathsf{R}_4-Y\cdot\mathsf{R}_3$	R_5	R_5	$2\mathsf{R}_6$	0	0	0
R_6	$-\mathsf{R}_5$	0	$2\mathsf{R}_6$	0	0	0	0

(2.37)

The $\mathbb{C}$-vector space generated by the vector fields R_i form a one-parameter family of seven-dimensional Lie algebras. Our main motivation for introducing such vector fields is that they are basic ingredients for the Bershadsky-Cecotti-Ooguri-Vafa holomorphic anomaly equation. This will be explained in the next section.

2.14 Gromov-Witten invariants

The present text only deals with the B-model Calabi-Yau variety of mirror symmetry. We can compute Gromov-Witten, Donaldson-Thomas and Gopakumar-Vafa invariants which has to do with the enumerative geometry of the A-model Calabi-Yau variety. It is beyond the scope of the present text to study these numbers. We refer the reader to [HKK$^+$03] and [MNOP06] on this topic. Let $N_{g,d}$ be the genus g Gromov-Witten invariants of the generic quintic in $\mathbb{P}^4$. The genus zero Gromov-Witten invariants are $N_{0,d} = \sum_{k|d} \frac{n_{d/k}}{k^3}$ which are gathered into the generating function F_0 called the prepotential:

$$F_0 = \frac{-200}{2}\frac{\zeta(3)}{(2\pi i)^3} + \frac{5}{12}\tau_0\left(\frac{5}{2} + \tau_0^2\right) + H,$$

where H is given in (2.35), see [KKV99, vEvS06]. Following the physics literature we call

$$\mathsf{F}_g^{\mathrm{hol}} := \sum_{d=0}^{\infty} N_{g,d}q^d,\ g \geq 2,\quad \mathsf{F}_1^{\mathrm{hol}} := \frac{25}{12}\ln q + \sum_{d=1}^{\infty} N_{1,d}q^d$$

the genus g topological string partition function. A discussion of such functions in the case of elliptic curves is done in §6.8. Computations of [BCOV93] leads us to the fact that $\mathsf{F}_1^{\mathrm{hol}}$ is obtained from the following algebraic object after doing q-expansion as in §2.8:

$$\mathsf{F}_1^{\mathrm{alg}} := \ln\left(t_4^{\frac{25}{12}}(t_4 - t_0^5)^{\frac{-5}{12}} t_5^{\frac{1}{2}}\right). \tag{2.38}$$

In a similar way, we can describe regular functions $\mathsf{F}_g^{\mathrm{alg}} \in \mathscr{O}_\mathsf{T}$ with the holomorphic counterpart $\mathsf{F}_g^{\mathrm{hol}}$. This is as follows. The Bershadsky-Cecotti-Ooguri-Vafa holomorphic anomaly equation for $\mathsf{F}_g^{\mathrm{alg}}$ turns out to be the following set of equations:

$$\begin{aligned} \mathsf{R}_i\mathsf{F}_g^{\mathrm{alg}} &= 0,\quad i = 1,3, \\ \mathsf{R}_2\mathsf{F}_g^{\mathrm{alg}} &= (2g-2)\mathsf{F}_g^{\mathrm{alg}}, \\ \mathsf{R}_4\mathsf{F}_g^{\mathrm{alg}} &= \frac{1}{2}\left(\mathsf{R}_0^2\mathsf{F}_{g-1}^{\mathrm{alg}} + \sum_{r=1}^{g-1} \mathsf{R}_0\mathsf{F}_r^{\mathrm{alg}}\mathsf{R}_0\mathsf{F}_{g-r}^{\mathrm{alg}}\right). \end{aligned} \tag{2.39}$$

These collections of equations do not determine $\mathsf{F}_g^{\mathrm{alg}}$ uniquely. It is defined up to addition by an element in $\mathscr{O}_\mathsf{S} := \mathsf{k}[t_0, t_4, \frac{1}{t_4 - t_0^5}]$. Using further Physics ansatz we have

Proposition 1 *For* $\mathsf{F}_g^{\mathrm{alg}}$, $g \geq 2$, *we have*

$$\mathsf{F}_g^{\mathrm{alg}} = \frac{Q_g}{(t_4 - t_0^5)^{2g-2} t_5^{3g-3}}, \tag{2.40}$$

where Q_g is a homogeneous polynomial of degree $69(g-1)$ with weights (2.11) and with rational coefficients, and so $\mathsf{F}_g^{\mathrm{alg}}$ is of degree $6g-6$. Further, any monomial $t_0^{i_0} t_1^{i_1} \cdots t_6^{i_6}$ in $\mathsf{F}_g^{\mathrm{alg}}$ satisfies $i_2+i_3+i_4+i_5+i_6 \geq 3g-3$.

One can also get the above statement by comparing period expressions for propagators in [YY04] and for t_i's in Theorem 4. For the discussion of the ambiguity part of $\mathsf{F}_g^{\mathrm{alg}}$ see §6.6. For the discussion of the enumerative meaning of the coefficients of $\mathsf{F}_g^{\mathrm{alg}}$ see [BCOV94] in the physics framework or [Kon95] in a mathematical framework. For further details see §6.

2.15 Periods and differential Calabi-Yau modular forms, II

Many arguments of the previous sections work for a huge class of fourth-order differential equations. Since the appearance of the Picard-Fuchs equation (2.25) there have been many efforts in mathematics to classify such differential equations which arise from the variations of Calabi-Yau varieties, see [GAZ10]. Therefore, it fits well into the context of the present text to develop differential Calabi-Yau modular forms for all these equations. In this section we give a summary of our computations in this direction.

We start with a fourth-order Fuchsian differential equation

$$\begin{aligned} L &:= P_0(\theta) - \sum_{i=1}^{k} z^i P_i(\theta) \\ &= a(z)\left(\theta^4 - \sum_{i=0}^{3} a_i(z)\theta^i\right) = 0, \end{aligned} \tag{2.41}$$

where $\theta = z\frac{\partial}{\partial z}$ and P_i's are polynomials of degree at most 4 and so a_i's are rational functions in z. We define:

$$a_4 := \frac{1}{4}a_3^2 + a_2 - \frac{1}{2}\theta a_3, \quad a_5(z) := e^{\frac{1}{2}\int a_3(z)\frac{dz}{z}}. \tag{2.42}$$

The reader may follow the computation of the present text for hypergeometric linear differential equations

$$L = \theta^4 - zP_1(\theta), \quad P_1(\theta) := (\theta^4 + r_3\theta^3 + r_2\theta^2 + r_1\theta + r_0) \tag{2.43}$$

for some constants $r_0, r_1, r_2, r_3 \in \mathbb{C}$. We are mainly interested on those linear differential equations $L = 0$ which come from geometry in the following sense, however, the main results of this chapter do not assume this. We consider a one-parameter family of threefolds X_z together with a homomorphic differential 3-form η. We further assume that $H^3(X_z,\mathbb{C})$ is of dimension 4 and its Hodge numbers are all one, $h^{30}=h^{21}=h^{12}=h^{03}=1$. The periods $\int_{\delta_z}\eta$, $\delta_z\in H_3(X_z,\mathbb{Z})$ satisfy a fourth-order linear differential equation $L=0$, which is usually called a Picard-Fuchs equation. In this case a_5 is a rational function in z, see §7.3. We have a well-known 14 examples of families of Calabi-Yau threefolds with the hypergeometric Picard-Fuchs equation, see [DM06]. The most well-known one is given by $P_1(\theta)=\prod_{i=1}^4(\theta+\frac{i}{5})$, which is (2.25). We refer to this as the number one Calabi-Yau equation, see [GAZ10].

After our computations in Theorem 4 and its modification in Theorem 10 we formulate the following definition. Let ψ_0 and ψ_1 be two solutions of $L=0$. The field of meromorphic differential Calabi-Yau modular forms is simply

$$\mathscr{O}_{\mathsf{T}}:=\mathbb{Q}(u_0,u_1,\ldots,u_6)$$

generated with seven quantities

$$\begin{aligned}
u_0 &:= z,\\
u_i &:= \theta^{i-1}\psi_0,\ i=1,2,3,4,\\
u_5 &:= \psi_0\theta\psi_1-\psi_1\theta\psi_0,\\
u_6 &:= \psi_0\theta^2\psi_1-\psi_1\theta^2\psi_0.
\end{aligned}$$

We assume a list of properties for L (see §7.1). For instance, we assume that there is a particular intersection form $\langle\cdot,\cdot\rangle$ in the solution space of $L=0$. We can write q-expansions for the elements of $\mathscr{O}_{\mathsf{T}}$ as follows. We assume that the indicial equation $P_0(\lambda)=0$ of $L=0$ at $z=0$ has a double root and so $L=0$ has two solutions of the form

$$\psi_0=1+O(z),\quad \psi_1=\psi_0\ln z+O(z).$$

All the elements of $\mathscr{O}_{\mathsf{T}}$ are holomorphic at $z=0$ and so they can be written in the new coordinate system

$$q=e^{\tau},\quad \text{where}\quad \tau:=\frac{\psi_1}{\psi_0}.$$

The function τ is called the mirror map in Physics literature. In this way, we get q-expansions for all elements of $\mathscr{O}_{\mathsf{T}}$. We assume that $\langle\psi_0,\psi_1\rangle=0$ and in this way $\mathscr{O}_{\mathsf{T}}$ is invariant under θ. We are mainly interested in the differential structure of $\mathscr{O}_{\mathsf{T}}$ given by

$$\dot{*}:=\frac{\partial *}{\partial\tau}=q\frac{\partial *}{\partial q}=\frac{u_1^2}{u_5}\theta(*).$$

We have

$$\begin{cases} \dot{u}_0 = \frac{u_1^2}{u_5} u_0, \\ \dot{u}_1 = \frac{u_1^2}{u_5} u_2, \\ \dot{u}_2 = \frac{u_1^2}{u_5} u_3, \\ \dot{u}_3 = \frac{u_1^2}{u_5} u_4, \\ \dot{u}_4 = \frac{u_1^2}{u_5} \sum_{i=0}^{3} a_i(u_0) u_{i+1}, \\ \dot{u}_5 = \frac{u_1^2}{u_5} u_6, \\ \dot{u}_6 = \frac{u_1^2}{u_5} \left(2\frac{u_2 u_6 - u_3 u_5}{u_1} + \frac{1}{2} a_3(u_0) u_6 + a_4(u_0) u_5 \right) \end{cases} \tag{2.44}$$

where a_i's are given in (2.41) and (2.42), see §7.7. The above differential equation can be considered as a generalization of Theorem 4 and the Ramanujan vector field satisfied by Eisenstein series.

The differential equation (2.44) computes the q-expansions of u_i's, and hence all the elements of $\mathscr{O}_\mathsf{T}$. We only need the initial conditions

$$\begin{aligned} u_0 &= 0 + q + O(q^2), \\ u_1 &= 1 + c_0 q + O(q^2), \\ u_2 &= 0 + c_0 q + O(q^2), \\ u_3 &= 0 + c_0 q + O(q^2), \\ u_4 &= 0 + c_0 q + O(q^2), \\ u_5 &= 1 + c_1 q + O(q^2), \\ u_6 &= 0 + c_1 q + O(q^2) \end{aligned} \tag{2.45}$$

where

$$c_0 := \frac{P_1(0)}{P_0(1)}, \quad c_1 := \frac{P_0(1)P_1'(0) - P_1(0)P_0'(1)}{P_0(1)^2} + 2\frac{P_1(0)}{P_0(1)}$$

and P_0 and P_1 are given in the expression of $L = 0$. If $P_0(1) = 0$, then we can take c_0 and c_1 arbitrary fixed numbers. Other coefficients of u_i's are given by a mixed recursion obtained from (2.44), see §7.9. For further details see §7.

In §8, we describe anti-holomorphic counterparts of differential Calabi-Yau modular forms. The topological partition functions are originally defined as anti-holomorphic functions and the corresponding anomaly equation involves anti-holomorphic derivations. In this chapter we explain how the translation from anti-holomorphic to holomorphic objects can be done. A similar topic in the case of elliptic curves and second-order differential equations is not covered in the literature and Appendix A fills this gap.

2.16 BCOV holomorphic anomaly equation, II

We would like to treat the Linear differential L as if it is associated to the variation of Calabi-Yau varieties. The good news is that we can regard $\mathscr{O}_{\mathsf{T}}$ as the field of meromorphic functions on the moduli of enhanced Calabi-Yau varieties, even though we do not know whether such geometric objects exists. We first recover the Gauss-Manin connection matrix A in this context. It is

$$\mathsf{A} := dg^{\mathsf{tr}} \cdot g^{-\mathsf{tr}} + g^{\mathsf{tr}}(A \cdot dz) g^{-\mathsf{tr}} \tag{2.46}$$

where g is given by

$$g = \begin{pmatrix} \frac{1}{u_1} & -\frac{u_2}{u_5} & -\frac{u_2u_6-u_3u_5}{a_5u_1^2} & -\frac{u_4-\frac{1}{3}a_3u_3-a_4u_2}{a_5} \\ 0 & \frac{u_1}{u_5} & \frac{u_6}{a_5u_1} & \frac{u_3-a_4u_1}{a_5} \\ 0 & 0 & -\frac{u_5}{a_5u_1} & -\frac{u_2+\frac{1}{3}a_3u_1}{a_5} \\ 0 & 0 & 0 & \frac{u_1}{a_5} \end{pmatrix}, \tag{2.47}$$

see §9.7. Now, it is a mere computation to verify Theorem 8 with this Gauss-Manin connection matrix. It turns out that the Yukawa coupling Y is given by $\mathsf{Y} := a_5(u_0) \cdot u_1^4/u_5^3$, see also §7.8, and the expression of R_i's in u_i's is given by

$$\begin{aligned}
\mathsf{R}_0 := {} & (-\frac{u_1^2}{u_5}u_0)\frac{\partial}{\partial u_0} + (-\frac{u_1^2}{u_5}u_2)\frac{\partial}{\partial u_1} + (-\frac{u_1^2}{u_5}u_3)\frac{\partial}{\partial u_2} + (-\frac{u_1^2}{u_5}u_4)\frac{\partial}{\partial u_3} \\
& + (-\frac{u_1^2}{u_5}\sum_{i=1}^{4} a_i(u_0)u_i)\frac{\partial}{\partial u_4} + (-\frac{u_1^2}{u_5}u_6)\frac{\partial}{\partial u_5} \\
& + (-\frac{u_1^2}{u_5}(2\frac{u_2u_6-u_3u_5}{u_1} + \frac{1}{2}a_3(u_0)u_6 + a_4(u_0)u_5)\frac{\partial}{\partial u_6}, \\
\mathsf{R}_1 = {} & u_5\frac{\partial}{\partial u_5} + u_6\frac{\partial}{\partial u_6}, \\
\mathsf{R}_2 = {} & u_1\frac{\partial}{\partial u_1} + u_2\frac{\partial}{\partial u_2} + u_3\frac{\partial}{\partial u_3} + u_4\frac{\partial}{\partial u_4} + u_5\frac{\partial}{\partial u_5} + u_6\frac{\partial}{\partial u_6}, \\
\mathsf{R}_3 = {} & \frac{u_5}{u_1}\frac{\partial}{\partial u_2} + \frac{u_6}{u_1}\frac{\partial}{\partial u_3} + \frac{u_2u_6 - u_3u_5 + \frac{1}{3}a_3u_1u_6 + a_4u_1u_5}{u_1^2}\frac{\partial}{\partial u_4}, \\
\mathsf{R}_4 = {} & \frac{a_5u_1^2}{u_5}\frac{\partial}{\partial u_6}, \\
\mathsf{R}_5 = {} & \frac{a_5u_1}{u_5}\frac{\partial}{\partial u_3} + \frac{a_5u_2 + \frac{1}{3}a_3a_5u_1}{u_5}\frac{\partial}{\partial u_4}, \\
\mathsf{R}_6 = {} & \frac{-a_5}{u_1}\frac{\partial}{\partial u_4}.
\end{aligned}$$

The proof is again based on explicit computations [Supp Item 3]. The vector fields R_i can be derived from the process of breaking an anti-holomorphic derivation into holomorphic ones. This is explained in §8. The Lie bracket between the vector field R_i's are given by (2.37). The BCOV holomorphic anomaly equation is the same as in (2.39). $\mathsf{F}_g^{\text{hol}}$ is defined up to addition of terms $u_1^{2g-2} p_g$ which we call it the ambiguity of $\mathsf{F}_g^{\text{hol}}$. Here, p_g is some rational function in z and one can compute its pole orders. Therefore, p_g depends only on a finite number of coefficients depending on g, see §9.11. For instance, for the hypergeometric differential equation p_g is a polynomial of degree at most $3g-3$ in $X := 1-a_5$, see §9.11.

The logarithmic derivative of the topological partition function of genus one is in $\mathscr{O}_{\mathsf{T}}$. In fact, we have

$$\mathsf{F}_1^{\text{hol}} := \frac{1}{2}\ln(u_5 u_1^{1+h^{21}+\frac{\chi}{12}} f(u_0))$$

where

$$f(z) = z^{r_0}(1-\frac{b_1}{z})^{r_1}(1-\frac{b_2}{z})^{r_2}\cdots(1-\frac{b_r}{z})^{r_r}, \quad r_0, r_1, \cdots, r_r \in \mathbb{Q}, \tag{2.48}$$

and $b_0 = 0, b_1, \ldots, b_r, \infty$ are the singularities of L. Here $h^{21} = 1$ and χ are respectively the Hodge and Euler number of the Calabi-Yau threefold X, see §9.1. If we do not know the geometric origin of $L = 0$, χ may be considered as another ambiguity constant. The rational numbers r_i are more ambiguities which cannot be derived from $L = 0$.

2.17 The polynomial structure of partition functions

Let

$$R = \mathbb{C}[z, a_5, \frac{1}{z-b_i}, \; i = 0, 1, 2, \ldots, r].$$

An element of R has poles only in the singularities of L. By definition we have $\theta^j a_i \in R, \;\; i = 0,1,2,3,4,5, \quad j = 0,1,2,\ldots$. The ring R is a differential ring with the differential operator θ. Since a_5 might not be a rational function in z, we had to add it to R. The Yamaguchi-Yau elements of $\mathscr{O}_{\mathsf{T}}$ are defined in the following way:

$$\begin{aligned}
X &:= 1 - a_5, \\
U &:= \frac{u_2}{u_1}, \\
V_1 &:= \frac{u_6}{u_5}, \\
V_2 &:= \frac{u_3}{u_1} - \frac{u_6}{u_5}\frac{u_2}{u_1}, \\
V_3 &:= \frac{u_4}{u_1} - \frac{u_2}{u_1}\left(-\frac{u_3}{u_1} + \frac{u_2}{u_1}\frac{u_6}{u_5} - \frac{1}{2}a_3\frac{u_6}{u_5} - a_4\right).
\end{aligned}$$

Hopefully, the usage of X as above and for a Calabi-Yau threefold will not produce any confusion.

Theorem 9 *The genus g topological partition function* $\mathsf{F}_g^{\mathrm{hol}}, g \geq 2$ *are elements in* $\mathscr{O}_{\mathsf{T}}$ *and in fact*

$$\mathsf{F}_g^{\mathrm{hol}} = \frac{u_1^{2g-2}}{(1-X)^{g-1}} Q_g, \quad Q_g := P_g + p_g, \tag{2.49}$$

where P_g *is a polynomial of degree* $3g-3$ *in the ring*

$$R[V_1, V_2, V_3], \quad \deg(V_1) = 1, \ \deg(V_2) = 2, \ \deg(V_3) = 3.$$

The polynomial P_g *does not depend on* U *and it has no constant term, that is,* $P_g(0,0,0) = 0$. *Here,* $p_g \in R$ *is the ambiguity of* F_g.

For a proof and further details see §9.

2.18 Future developments

It is highly desirable to see the theory presented here as useful as modular forms. However, we are blocked with many open problems so that we do not see how the further analogies between our theory of modular forms and the classical theory will be established. In §11 we have collected a bunch of such problems and we hope that it will pave the road for further investigations. Here, is a brief description of §11. The lack of Hecke operators and Hecke algebras can be considered as one of the main issues for differential Calabi-Yau modular forms. We discuss this in both algebraic geometric and complex analysis context. Global properties of periods and mirror maps is also another main problem in our context. It can be reformulated in terms of vanishing of periods, maximal Hodge structures and global Torelli problem. The existence of a kind of monstrous moonshine conjecture in our context would create a fascination in mathematics. Despite many attempts, the integrality of instanton numbers, and many others arising from some product formulas, are not fully established. This will require a heavy machinery of p-adic analysis. We can choose different mirror maps and we still get integral q-expansion, or q-expansions with rational coefficients such that the primes appear in their denominators very s-lowly. Concrete statements supported by computer calculations will appear in our way. Around the conifold singularity no enumerative geometry is known, however, we have the gap condition for topological partitions functions which is not completely understood in mathematical terms. Finally, the last dream problem would be to see an example of a differential Calabi-Yau modular form used in the arithmetic modularity of a variety.

Chapter 3
Moduli of enhanced mirror quintics

This chapter is dedicated to algebraic-geometric aspects, such as moduli space and algebraic de Rham cohomology, of mirror quintics. We will use the algebraic de Rham cohomology $H^3_{\mathrm{dR}}(X)$ of X defined over an arbitrary field k of characteristic zero. The original text of Grothendieck [Gro66] is still the main source of information for algebraic de Rham cohomology. In the present text by the moduli of objects x we mean the set of all x modulo natural isomorphisms among them. The computer codes used in the present chapter can be found in the author's webpage [Supp Item 4].

3.1 What is mirror quintic?

Let k be a field of characteristic zero. A mirror quintic Calabi-Yau threefold, mirror quintic for short, over k is the variety $X = X_z$ obtained by a resolution of singularities of the quotient W_z/G, where W_z is the Dwork family of quintics in $\mathbb{P}^4$:

$$W_z := \{[x_0 : x_1 : x_2 : x_3 : x_4] \in \mathbb{P}^4 \mid z \cdot x_0^5 + x_1^5 + x_2^5 + x_3^5 + x_4^5 - 5x_0x_1x_2x_3x_4 = 0\}, \quad (3.1)$$

$$z \neq 0, 1, \ z \in \mathsf{k}$$

and G is the group

$$G := \{(\zeta_1, \zeta_2, \ldots, \zeta_5) \mid \zeta_i^5 = 1, \ \zeta_1\zeta_2\zeta_3\zeta_4\zeta_5 = 1\}$$

acting on W_z coordinatewise. Such a resolution is described explicitly in [Mor93], however, for the purpose of the present text we do not need it.

The moduli of mirror quintics is the punctured line $\mathbb{P}^1 - \{0, 1\}$ with the coordinate system z. Note that for $z = \infty$ we have still a smooth mirror quintic and so there is no universal family of mirror quintics over the punctured line $\mathbb{P}^1 - \{0, 1\}$.

Mirror quintic Calabi-Yau threefolds produced a great amount of excitement in mathematics when in 1991 Candelas et al. in [CdlOGP91] used its periods (in the

B-model of Topological String Theory) in order to compute the number of rational curves in a generic quintic (A-model of Topological String Theory). This was the first concrete evidence for the importance of mirror symmetry as a device which produces conjectures beyond the imagination of mathematicians.

Definition 1 For reasons that will be clear later, we call $z = 0$ the singularity with maximal unipotent monodromy (MUM singularity), and we call $z = 1$ the conifold singularity. Following the literature in modular forms we may use the word cusp instead of singularity.

3.2 Moduli space, I

We first construct explicit affine coordinates for the moduli S of pairs (X, ω), where X is a mirror quintic and ω is a holomorphic differential 3-form on X. We have

$$\mathsf{S} = \mathrm{Spec}(\mathbb{Q}[t_0, t_4, \frac{1}{(t_0^5 - t_4)t_4}]),$$

where for $(t_0, t_4) \in \mathsf{S}(\bar{\mathsf{k}})$ we associate the pair $(\mathsf{X}_{t_0,t_4}, \omega)$. In the affine coordinates (x_1, x_2, x_3, x_4), that is $x_0 = 1$, X_{t_0,t_4} is given by a desingularization of $\{f(x) = 0\}/G$, where

$$f(x) := -t_4 - x_1^5 - x_2^5 - x_3^5 - x_4^5 + 5t_0x_1x_2x_3x_4,$$

and

$$\omega := \frac{dx_1 \wedge dx_2 \wedge dx_3 \wedge dx_4}{df}.$$

Here, the finite group G acts on $f(x) = 0$ in the same way as in §3.1. The multiplicative group $\mathbb{G}_m := (\mathsf{k}^*, \cdot)$ acts on S by

$$(X, \omega) \bullet k = (X, k^{-1}\omega),\ k \in \mathbb{G}_m,\ (X, \omega) \in \mathsf{S}.$$

In the coordinates (t_0, t_4) this corresponds to

$$(t_0, t_4) \bullet k = (kt_0, k^5 t_4),\ (t_0, t_4) \in \mathsf{S},\ k \in \mathbb{G}_m. \tag{3.2}$$

This can be verified using the isomorphism

$$(\mathsf{X}_{(kt_0, k^5 t_4)}, k\omega) \cong (\mathsf{X}_{(t_0, t_4)}, \omega),\quad (x_1, x_2, x_3, x_4) \mapsto (k^{-1}x_1, k^{-1}x_2, k^{-1}x_3, k^{-1}x_4). \tag{3.3}$$

Two Calabi-Yau varieties in the family (3.1) are isomorphic if and only if they have the same z. This and (3.2) imply that distinct pairs (t_0, t_4) give non-isomorphic pairs (X, ω). As a consequence of our construction we have

Proposition 2 *The morphism* $\mathsf{X} \to \mathsf{S}$ *is the universal family of* (X, ω)*, where* X *is mirror quintic and* ω *is holomorphic 3-form on* X.

In Physics literature one usually use the pair $(\mathsf{X}_{\psi,1}, 5\psi\omega)$, see [CdlOGP91, Mor92].

Remark 1 In the projective space $\mathbb{P}^4$ one may define $\mathsf{X}_{t_0,a} : F = 0,\ \ F := a_0x_0^5 + a_1x_1^5 + a_2x_2^5 + a_3x_3^5 + a_4x_4^5 + 5t_0x_0x_1x_2x_3x_4 = 0$ and Ω to be the residue of $\frac{\sum_{i=0}^4 x_i \widehat{dx_i}}{F}$ in $\mathsf{X}_{t_0,a}$. The pair $(\mathsf{X}_{t_0,a}, \Omega)$ only depends on $(t_0, a_0a_1a_2a_3a_4)$ and this is the main reason why we can set $a_0 = t_4, a_1 = a_2 = a_3 = a_4 = 1$.

3.3 Gauss-Manin connection, I

For a proper smooth family X/T of algebraic varieties defined over a field k of characteristic zero, we have the Gauss-Manin connection

$$\nabla : H^i_{\mathrm{dR}}(\mathsf{X}/\mathsf{T}) \to \Omega^1_{\mathsf{T}} \otimes_{\mathscr{O}_{\mathsf{T}}} H^i_{\mathrm{dR}}(\mathsf{X}/\mathsf{T}),$$

where $H^i_{\mathrm{dR}}(\mathsf{X}/\mathsf{T})$ is the i-th relative de Rham cohomology and Ω^1_{T} is the set of differential 1-forms on T. For simplicity we have assumed that T is affine and $H^i_{\mathrm{dR}}(\mathsf{X}/\mathsf{T})$ is an $\mathscr{O}_{\mathsf{T}}$-module, where $\mathscr{O}_{\mathsf{T}}$ is the k-algebra of regular function on T. By definition of a connection, ∇ is k-linear and satisfies the Leibniz rule

$$\nabla(r\omega) = dr \otimes \omega + r\nabla\omega, \omega \in H^i_{\mathrm{dR}}(\mathsf{X}/\mathsf{T}),\ r \in \mathscr{O}_{\mathsf{T}}.$$

For a vector field v in T we define

$$\nabla_v : H^i_{\mathrm{dR}}(\mathsf{X}/\mathsf{T}) \to H^i_{\mathrm{dR}}(\mathsf{X}/\mathsf{T})$$

to be ∇ composed with

$$v \otimes \mathrm{Id} : \Omega^1_{\mathsf{T}} \otimes_{\mathscr{O}_{\mathsf{T}}} H^i_{\mathrm{dR}}(\mathsf{X}/\mathsf{T}) \to \mathscr{O}_{\mathsf{T}} \otimes_{\mathscr{O}_{\mathsf{T}}} H^i_{\mathrm{dR}}(\mathsf{X}/\mathsf{T}) = H^i_{\mathrm{dR}}(\mathsf{X}/\mathsf{T}).$$

Sometimes it is useful to choose a basis $\omega_1, \omega_2, \ldots, \omega_h$ of the $\mathscr{O}_{\mathsf{T}}$-module $H^i(X/\mathsf{T})$ and write the Gauss-Manin connection in this basis:

$$\nabla \begin{pmatrix} \omega_1 \\ \omega_2 \\ \vdots \\ \omega_h \end{pmatrix} = \mathsf{A} \otimes \begin{pmatrix} \omega_1 \\ \omega_2 \\ \vdots \\ \omega_h \end{pmatrix}, \tag{3.4}$$

where A is an $h \times h$ matrix with entries in Ω^1_{T}. For further information on Gauss-Manin connection see [KO68]. See also [Mov11b] for computational aspects of Gauss-Manin connection.

3.4 Intersection form and Hodge filtration

For $\omega,\alpha \in H^3_{\mathrm{dR}}(X)$ let

$$\langle \omega,\alpha\rangle := \mathrm{Tr}(\omega\cup\alpha)$$

be the intersection form in $H^3_{\mathrm{dR}}(X)$. If X is defined over complex numbers, then we can consider it as a complex manifold and its de Rham cohomology defined by C^∞ forms, and the intersection form is just

$$\langle \omega,\alpha\rangle = \frac{1}{(2\pi i)^3}\int_{X(\mathbb{C})} \omega\wedge\alpha.$$

For all these see for instance Deligne's lectures in [DMOS82]. In $H^3_{\mathrm{dR}}(X)$ we have the Hodge filtration

$$\{0\} = F^4 \subset F^3 \subset F^2 \subset F^1 \subset F^0 = H^3_{\mathrm{dR}}(X),\ \ \dim_{\mathbb{C}}(F^i) = 4-i.$$

There is a relation between the Hodge filtration and the intersection form which is given by the following collection of equalities:

$$\langle F^i, F^j\rangle = 0,\ i+j\geq 4.$$

Now let us consider the universal family X/S constructed in §3.2. The Griffiths transversality is a property combining the Gauss-Manin connection and the Hodge filtration. It says that the Gauss-Manin connection sends F^i to $\Omega^1_{\mathsf{S}}\otimes F^{i-1}$ for $i = 1,2,3$. Using this we conclude that:

$$\omega_i := \frac{\partial^{i-1}}{\partial t_0^{i-1}}(\omega) \in F^{4-i},\ i = 1,2,3,4. \tag{3.5}$$

By abuse of notation we have used $\frac{\partial}{\partial t_0}$ instead of $\nabla_{\frac{\partial}{\partial t_0}}$.

Proposition 3 *The intersection form in the basis ω_i is*

$$[\langle \omega_i,\omega_j\rangle] = \begin{pmatrix} 0 & 0 & 0 & \frac{1}{625}(t_4-t_0^5)^{-1} \\ 0 & 0 & -\frac{1}{625}(t_4-t_0^5)^{-1} & -\frac{1}{125}t_0^4(t_4-t_0^5)^{-2} \\ 0 & \frac{1}{625}(t_4-t_0^5)^{-1} & 0 & \frac{1}{125}t_0^3(t_4-t_0^5)^{-2} \\ -\frac{1}{625}(t_4-t_0^5)^{-1} & \frac{1}{125}t_0^4(t_4-t_0^5)^{-2} & -\frac{1}{125}t_0^3(t_4-t_0^5)^{-2} & 0 \end{pmatrix}.$$

We will prove this in §3.8.

3.5 A vector field on S

In (3.5) we have chosen the vector field $\frac{\partial}{\partial t_0}$ in S in order to produce the basis $\omega_i,\ i = 1,2,3,4$. In this section we would like to give an intrinsic characterization of such a vector field with respect to the moduli S.

Let v be the vector field on S which produces the $\mathbb{G}_m$-action. This is defined as follows. For $x \in \mathsf{S}$ we define $f_x : \mathbb{A}^1_{\mathsf{k}} - \{0\} \to \mathsf{S},\ f_x(a) := x \bullet a$. It maps the point 1 to x. The vector field v at the point x is just the image of the vector 1 over the point $1 \in \mathbb{A}^1_{\mathsf{k}}$ and under the derivation of f_x. It is easy to show that this vector field in the case of elliptic curves and mirror quintics is

$$2t_2\frac{\partial}{\partial t_2} + 3t_3\frac{\partial}{\partial t_3} \text{ and } t_0\frac{\partial}{\partial t_0} + 5t_4\frac{\partial}{\partial t_4},$$

respectively. We are looking for a vector field w in S such that it is orthogonal to v in all the points of S. In the case of elliptic curves we have

$$w = 9t_3\frac{\partial}{\partial t_2} + \frac{1}{2}t_2\frac{\partial}{\partial t_3}$$

because

$$w \wedge v = (27t_3^2 - t_2^3)\frac{\partial}{\partial t_2} \wedge \frac{\partial}{\partial t_3}.$$

In the case of mirror quintics we have two possibilities for w which are due to two discriminant loci $t_4 = 0,\ t_4 - t_0^5 = 0$:

$$w := \frac{1}{5}\frac{\partial}{\partial t_0},\ \frac{1}{5}\frac{\partial}{\partial t_0} + t_0^4\frac{\partial}{\partial t_4}. \tag{3.6}$$

The existence of the vector field w is essential for constructing explicit affine coordinates for the moduli T introduced in the next section. Another important property that we would like to have is that $\langle \omega, \nabla_w^3 \omega \rangle$ is invertible in $\mathscr{O}_{\mathsf{S}}$, and hence, $\nabla_w^i \omega,\ i = 0,1,\ldots,3$ form a basis of $H^3_{\mathrm{dR}}(\mathsf{X}_t)$ for all $\mathsf{t} \in \mathsf{S}$. In the mirror quintic case with the first vector field in (3.6), this follows from the $(1,4)$ entry of the matrix in Proposition 3.

3.6 Moduli space, II

We make the base change $\alpha = S\omega$, where S is given by

$$S = \begin{pmatrix} 1 & 0 & 0 & 0 \\ t_9 & t_8 & 0 & 0 \\ t_7 & t_6 & t_5 & 0 \\ t_1 & t_2 & t_3 & t_{10} \end{pmatrix} \tag{3.7}$$

and t_i's are unknown parameters, and we assume the intersection form in α_i's is given by the matrix Φ in (0.1):

$$\Phi = [\langle \alpha_i, \alpha_j \rangle] = S[\langle \omega_i, \omega_j \rangle]S^{\mathrm{tr}}.$$

This yields to many polynomial relations between t_i's. It turns out that we can take t_i, $i = 0, 1, 2, 3, \ldots, 6$ as independent parameters and calculate all others in terms of these seven parameters:

$$\begin{aligned} t_7t_8 - t_6t_9 &= 3125t_0^3 + t_2, \\ t_{10} &= -t_8t_5, \\ t_5t_9 &= -3125t_0^4 - t_3, \\ t_{10} &= 625(t_4 - t_0^5). \end{aligned}$$

The expression of t_7, t_8, t_9 are given in (2.12). The moduli space T introduced in Theorem 2 turns out to be the one in (2.8). Note that we have

$$\mathsf{T}(\bar{\mathsf{k}}) \cong \{(t_0, t_1, t_2, t_3, t_4, t_5, t_6) \in \bar{\mathsf{k}}^7 \mid t_5t_4(t_4 - t_0^5) \neq 0\}. \tag{3.8}$$

Here, for $t \in \mathsf{T}(\bar{\mathsf{k}})$ in the right hand side of the isomorphism (3.8), we associate the pair $(\mathsf{X}_{t_0,t_4}, \alpha)$. By our construction we have a morphism $\mathsf{X} \to \mathsf{T}$ of algebraic varieties over k and elements $\alpha_i \in H^3_{\mathrm{dR}}(\mathsf{X}/\mathsf{T}),\ \ i = 1, 2, 3, 4$. We conclude that

Proposition 4 *The morphism $\mathsf{X} \to \mathsf{T}$ is the universal family of (X, α), where X is a mirror quintic and α is a basis of the third algebraic de Rham cohomology $H^3_{\mathrm{dR}}(X)$, compatible with the Hodge filtration and with the constant intersection matrix Φ.*

Remark 2 We also define

$$\tilde{t}_5 = \frac{1}{3125}t_5, \quad \tilde{t}_6 = -\frac{1}{5^6}(t_0^5 - t_4)t_6 + \frac{1}{5^{10}}(9375t_0^4 + 2t_3)t_5$$

which correspond to the parameters t_5, t_6 in the previous article [Mov11a].

3.7 The Picard-Fuchs equation

Let us consider the one-parameter family of Calabi-Yau varieties X_{t_0,t_4} with $t_0 = 1$ and $t_4 = z$ and denote by η the restriction of ω_1 to these parameters. We calculate the Picard-Fuchs equation of η with respect to the parameter z:

$$\frac{\partial^4 \eta}{\partial z^4} = \sum_{i=1}^{4} a_i(z) \frac{\partial^{i-1} \eta}{\partial z^{i-1}} \quad \text{modulo relatively exact forms.}$$

This is in fact the linear differential equation

$$I'''' = \frac{-24}{625z^3(z-1)}I + \frac{-24z+5}{5z^3(z-1)}I' + \frac{-72z+35}{5z^2(z-1)}I'' + \frac{-8z+6}{z(z-1)}I''', \quad ' = \frac{\partial}{\partial z} \tag{3.9}$$

which is calculated in [CdlOGP91]. This differential equation can be also written in the form (2.25). In [Mov11b] we have developed algorithms which calculate such differential equations. The parameter z is more convenient for our calculations than the parameter ψ used in Physics literature. The differential equation (3.9) is satisfied by the periods

$$I(z) = \int_{\delta_z} \eta, \ \delta \in H_3(\mathsf{X}_{1,z}, \mathbb{Q})$$

of the differential form η on the family $\mathsf{X}_{1,z}$. In the basis $\frac{\partial^i \eta}{\partial z^i}, \quad i = 0,1,2,3$ of $H^3_{\rm dR}(\mathsf{X}_{1,z})$ the Gauss-Manin connection matrix has the form

$$A(z)dz := \begin{pmatrix} 0 & 1 & 0 & 0 \\ 0 & 0 & 1 & 0 \\ 0 & 0 & 0 & 1 \\ a_1(z) & a_2(z) & a_3(z) & a_4(z) \end{pmatrix} dz. \tag{3.10}$$

3.8 Gauss-Manin connection, II

We would like to calculate the Gauss-Manin connection

$$\nabla : H^3_{\rm dR}(\mathsf{X}/\mathsf{S}) \to \Omega^1_{\mathsf{S}} \otimes_{\mathscr{O}_{\mathsf{S}}} H^3_{\rm dR}(\mathsf{X}/\mathsf{S})$$

of the two parameter proper family of varieties X_{t_0,t_4}, $(t_0,t_4) \in S$. We calculate ∇ with respect to the basis (3.5) of $H^3_{\rm dR}(\mathsf{X}/S)$. For this purpose we return back to the one parameter case. For $z := \frac{t_4}{t_0^5}$, consider the map

$$g : \mathsf{X}_{(t_0,t_4)} \to \mathsf{X}_{1,z},$$

given by (3.3) with $k = t_0^{-1}$. We have $g^*\eta = t_0\omega_1$ and

$$\frac{\partial}{\partial z} = \frac{-1}{5}\frac{t_0^6}{t_4}\frac{\partial}{\partial t_0} \left(= t_0^5 \frac{\partial}{\partial t_4} \right).$$

From these two equalities we obtain a matrix $\tilde{S} = \tilde{S}(t_0,t_4)$ such that

$$[\eta, \frac{\partial \eta}{\partial z}, \frac{\partial^2 \eta}{\partial z^2}, \frac{\partial^3 \eta}{\partial z^3}]^{\mathsf{tr}} = \tilde{S}^{-1}[\omega_1, \omega_2, \omega_3, \omega_4]^{\mathsf{tr}},$$

where tr denotes the transpose of matrices, and the Gauss-Manin connection in the basis ω_i, $i = 1,2,3,4$ is

$$\tilde{\mathsf{A}} = \left(d\tilde{S} + \tilde{S} \cdot A(\frac{t_4}{t_0^5}) \cdot d(\frac{t_4}{t_0^5}) \right) \cdot \tilde{S}^{-1},$$

which is the following matrix after doing explicit calculations:

$$\begin{pmatrix} -\frac{1}{5t_4} dt_4 & dt_0 + \frac{-t_0}{5t_4} dt_4 & 0 & 0 \\ 0 & \frac{-2}{5t_4} dt_4 & dt_0 + \frac{-t_0}{5t_4} dt_4 & 0 \\ 0 & 0 & \frac{-3}{5t_4} dt_4 & dt_0 + \frac{-t_0}{5t_4} dt_4 \\ \frac{-t_0}{t_0^5 - t_4} dt_0 + \frac{t_0^2}{5t_0^5 t_4 - 5t_4^2} dt_4 & \frac{-15t_0^2}{t_0^5 - t_4} dt_0 + \frac{3t_0^3}{t_0^5 t_4 - t_4^2} dt_4 & \frac{-25t_0^3}{t_0^5 - t_4} dt_0 + \frac{5t_0^4}{t_0^5 t_4 - t_4^2} dt_4 & \frac{-10t_0^4}{t_0^5 - t_4} dt_0 + \frac{6t_0^5 + 4t_4}{5t_0^5 t_4 - 5t_4^2} dt_4 \end{pmatrix}. \tag{3.11}$$

Now, we calculate the Gauss-Manin connection matrix of the family X/T written in the basis α_i, $i = 1,2,3,4$. This is

$$\mathsf{A} = \left(dS + S \cdot \tilde{\mathsf{A}} \right) \cdot S^{-1},$$

where S is the base change matrix (3.7). Since the matrix A is huge and does not fit into a mathematical paper, we do not write it here.

Proof of Proposition 3: Let Ω be the differential form ω with restricted parameters $t_0 = \psi$ and $t_4 = 1$. We have

$$\langle 5\psi\Omega, \frac{\partial^3(5\psi\Omega)}{\partial^3\psi} \rangle = \frac{1}{5^2} \frac{\psi^2}{1 - \psi^5},$$

see for instance [CdlOGP91], (4.6). From this we get

$$\langle \omega_1, \omega_4 \rangle = 5^{-4} \frac{1}{t_4 - t_0^5}. \tag{3.12}$$

We make the derivation of the equalities $\langle \omega_1, \omega_3 \rangle = 0$ and (3.12) with respect to t_0 and use the Picard-Fuchs equation of ω_1 with respect to the parameter t_0 and with t_4 fixed:

$$\frac{\partial \omega_4}{\partial t_0} = M_{41}\omega_1 + M_{42}\omega_2 + M_{43}\omega_3 + M_{44}\omega_4.$$

Here, M_{ij} is the (i,j)-entry of (3.11) after setting $dt_4 = 0$, $dt_0 = 1$. We get

$$\langle \omega_2, \omega_3 \rangle = -\langle \omega_1, \omega_4 \rangle, \ \langle \omega_2, \omega_4 \rangle = \frac{\partial \langle \omega_1, \omega_4 \rangle}{\partial t_0} - M_{44} \langle \omega_1, \omega_4 \rangle.$$

Derivating further the second equality we get

$$\langle \omega_3, \omega_4 \rangle = \frac{\partial \langle \omega_2, \omega_4 \rangle}{\partial t_0} - M_{43} \langle \omega_2, \omega_3 \rangle - M_{44} \langle \omega_2, \omega_4 \rangle.$$

3.9 Proof of Theorem 2

We are in the final step of the proof of Theorem 2. We have calculated the Gauss-Manin connection A written in the basis $\alpha_i,\ i=1,2,3,4$. It is a matter of explicit linear algebra calculations to show that there is a unique vector field R in T with the properties mentioned in Theorem 2 and to calculate it. In summary, the Gauss-Manin connection matrix composed with the vector field R and written in the basis α_i has the form

$$\nabla_{\mathsf{R}} = \begin{pmatrix} 0 & 1 & 0 & 0 \\ 0 & 0 & \frac{5^8(t_4-t_0^5)^2}{t_5^3} & 0 \\ 0 & 0 & 0 & -1 \\ 0 & 0 & 0 & 0 \end{pmatrix}. \tag{3.13}$$

It is interesting that the Yukawa coupling appears as the only non-constant term in the above matrix.

3.10 Algebraic group

There is an algebraic group which acts on the right hand side of the isomorphism (2.8). It corresponds to the base change in $\alpha_i,\ i=1,2,3,4$ such that the new basis is still compatible with the Hodge filtration and we have still the intersection matrix (0.1):

$$\mathsf{G} := \{\mathsf{g} = [\mathsf{g}_{ij}]_{4\times 4} \in \mathrm{GL}(4,\mathsf{k}) \mid \mathsf{g}_{ij}=0,\ \text{ for } j<i \text{ and } \mathsf{g}^{\mathrm{tr}}\Phi\mathsf{g}=\Phi\},$$

$$\left\{ \mathsf{g} = \begin{pmatrix} \mathsf{g}_{11} & \mathsf{g}_{12} & \mathsf{g}_{13} & \mathsf{g}_{14} \\ 0 & \mathsf{g}_{22} & \mathsf{g}_{23} & \mathsf{g}_{24} \\ 0 & 0 & \mathsf{g}_{33} & \mathsf{g}_{34} \\ 0 & 0 & 0 & \mathsf{g}_{44} \end{pmatrix},\ \mathsf{g}_{ij}\in\mathbb{C},\quad \begin{matrix} \mathsf{g}_{11}\mathsf{g}_{44}=1, \\ \mathsf{g}_{22}\mathsf{g}_{33}=1, \\ \mathsf{g}_{12}\mathsf{g}_{44}+\mathsf{g}_{22}\mathsf{g}_{34}=0, \\ \mathsf{g}_{13}\mathsf{g}_{44}+\mathsf{g}_{23}\mathsf{g}_{34}-\mathsf{g}_{24}\mathsf{g}_{33}=0 \end{matrix} \right\}.$$

G is called the Borel subgroup of $\mathrm{Sp}(4,\mathbb{C})$ respecting the Hodge flag. The action of G on the moduli T is given by

$$(X,[\alpha_1,\alpha_2,\alpha_3,\alpha_4])\bullet\mathsf{g} = (X,[\alpha_1,\alpha_2,\alpha_3,\alpha_4]\mathsf{g}).$$

The algebraic group G is of dimension six and has two multiplicative subgroup $\mathbb{G}_m=(\mathsf{k}^*,\cdot)$ and four additive subgroup $\mathbb{G}_a=(\mathsf{k},+)$ which generate it. In fact, an element $\mathsf{g}\in\mathsf{G}$ can be written in a unique way as the following product:

$$\begin{pmatrix} \mathsf{g}_1^{-1} & -\mathsf{g}_3\mathsf{g}_1^{-1} & (-\mathsf{g}_3\mathsf{g}_6+\mathsf{g}_4)\mathsf{g}_1^{-1} & (-\mathsf{g}_3\mathsf{g}_4+\mathsf{g}_5)\mathsf{g}_1^{-1} \\ 0 & \mathsf{g}_2^{-1} & \mathsf{g}_6\mathsf{g}_2^{-1} & \mathsf{g}_4\mathsf{g}_2^{-1} \\ 0 & 0 & \mathsf{g}_2 & \mathsf{g}_2\mathsf{g}_3 \\ 0 & 0 & 0 & \mathsf{g}_1 \end{pmatrix}$$

$$= \begin{pmatrix} g_1^{-1} & 0 & 0 & 0 \\ 0 & 1 & 0 & 0 \\ 0 & 0 & 1 & 0 \\ 0 & 0 & 0 & g_1 \end{pmatrix} \begin{pmatrix} 1 & 0 & 0 & 0 \\ 0 & g_2^{-1} & 0 & 0 \\ 0 & 0 & g_2 & 0 \\ 0 & 0 & 0 & 1 \end{pmatrix} \begin{pmatrix} 1 & -g_3 & 0 & 0 \\ 0 & 1 & 0 & 0 \\ 0 & 0 & 1 & g_3 \\ 0 & 0 & 0 & 1 \end{pmatrix} \begin{pmatrix} 1 & 0 & g_4 & 0 \\ 0 & 1 & 0 & g_4 \\ 0 & 0 & 1 & 0 \\ 0 & 0 & 0 & 1 \end{pmatrix} \begin{pmatrix} 1 & 0 & 0 & g_5 \\ 0 & 1 & 0 & 0 \\ 0 & 0 & 1 & 0 \\ 0 & 0 & 0 & 1 \end{pmatrix} \begin{pmatrix} 1 & 0 & 0 & 0 \\ 0 & 1 & g_6 & 0 \\ 0 & 0 & 1 & 0 \\ 0 & 0 & 0 & 1 \end{pmatrix}.$$

In other words, we have a bijection of sets $\mathbb{G}_m \times \mathbb{G}_m \times \mathbb{G}_a \times \mathbb{G}_a \times \mathbb{G}_a \times \mathbb{G}_a \cong \mathsf{G}$ sending $(g_i)_{i=1,\dots,6}$ to the above product. If we identify an element $g \in \mathsf{G}$ with the vector $(g_i)_{i=1,\dots,6}$ then

$$(g_1, g_2, g_3, g_4, g_5, g_6)^{-1}$$
$$= (g_1^{-1}, g_2^{-1}, -g_1^{-1} g_2 g_3, g_1^{-1} g_2^{-1}(g_3 g_6 - g_4), g_1^{-2}(-g_3^2 g_6 + 2 g_3 g_4 - g_5), -g_2^{-2} g_6).$$

We denote by $\bullet$ the right action of G on the T.

Proposition 5 *The action of G on t_i (as regular functions on the affine variety T) is given by*

$$\begin{aligned} g \bullet t_0 &= t_0 g_1, \\ g \bullet t_1 &= t_1 g_1^2 + t_7 g_1 g_2 g_3 + t_9 g_1 g_2^{-1} g_4 - g_3 g_4 + g_5, \\ g \bullet t_2 &= t_2 g_1^3 + t_6 g_1^2 g_2 g_3 + t_8 g_1^2 g_2^{-1} g_4, \\ g \bullet t_3 &= t_3 g_1^4 + t_5 g_1^3 g_2 g_3, \\ g \bullet t_4 &= t_4 g_1^5, \\ g \bullet t_5 &= t_5 g_1^3 g_2, \\ g \bullet t_6 &= t_6 g_1^2 g_2 + t_8 g_1^2 g_2^{-1} g_6. \end{aligned} \tag{3.14}$$

Consequently

$$\begin{aligned} g \bullet t_7 &= t_7 g_1 g_2 + t_9 g_1 g_2^{-1} g_6 - g_3 g_6 + g_4, \\ g \bullet t_8 &= t_8 g_1^2 g_2^{-1}, \\ g \bullet t_9 &= t_9 g_1 g_2^{-1} - g_3, \\ g \bullet t_{10} &= t_{10} g_1^5. \end{aligned}$$

Proof. We first calculate the action of $g = (k, 1, 0, 0, 0, 0)$, $k \in \mathsf{k}^*$ on t. We have an isomorphism $(\mathsf{X}_{(t_0,t_4)}, k^{-1}\omega_1) \cong (\mathsf{X}_{(t_0 k, t_4 k^5)}, \omega_1)$ given by

$$(x_1, x_2, x_3, x_4) \mapsto (k^{-1}x_1, k^{-1}x_2, k^{-1}x_3, k^{-1}x_4).$$

Under this isomorphism the vector field $k^{-1}\frac{\partial}{\partial t_0}$ is mapped to $\frac{\partial}{\partial t_0}$ and so $k^{-i}\omega_i$ is mapped to ω_i. This implies the isomorphisms

$$(\mathsf{X}_{(t_0,t_4)}, k(t_1\omega_1 + t_2\omega_2 + t_3\omega_3 + 625(t_4 - t_0^5)\omega_4))$$

$$\cong (\mathsf{X}_{(t_0,t_4)\bullet k}, k^2 t_1\omega_1 + k^3 t_2\omega_2 + k^4 t_3\omega_3 + 625(k^5 t_4 - (k t_0)^5)\omega_4))$$

and

$$(\mathsf{X}_{(t_0,t_4)}, S\omega) \cong (\mathsf{X}_{(t_0,t_4)\bullet k}, S\begin{pmatrix} k & 0 & 0 \\ 0 & k^2 & 0 & 0 \\ 0 & 0 & k^3 & 0 \\ 0 & 0 & 0 & k^4 \end{pmatrix}\omega),$$

where S is defined in (3.7). Therefore,

$$g \bullet t_i = t_i k^{\tilde{d}_i},\ \tilde{d}_i = i+1,\ i = 0,1,2,3,4\quad \tilde{d}_5 = 3,\ \tilde{d}_6 = 2.$$

Remark 3 The choice of the coordinates t_0, t_4 is unique (up to a multiplication by constants) and this follows from their functional equations in (3.14) and holomorphicity at cusps which we will describe it in §10. However, the choice of other t_i's is not unique, for instance, we could use $t_1 + t_0^2$ instead of t_1 and so on. Recall that in the case of elliptic curves t_1, t_2, t_3, which give us the Eisenstein series, are uniquely determined by their functional equations and holomorphicity at cusps, see [Mov12b].

The loci $t_4 = 0$ and $t_4 - t_0^5 = 0$ in the partial compactification of T correspond to the MUM and conifold singularities, respectively. The locus $t_5 = 0$ do not correspond to degeneration of mirror quintics and it can be considered as a part of the discriminant locus responsible for degeneration of differential forms α_i. Such a locus do not appear in the case of elliptic curves.

3.11 Another vector field

We are interested in a vector field $\check{\mathsf{R}}_0$ in T such that

$$\begin{aligned}
\nabla_{\check{\mathsf{R}}_0}(\alpha_1) &= \alpha_2, \\
\nabla_{\check{\mathsf{R}}_0}(\alpha_2) &= \check{\mathsf{Y}}\alpha_3, \\
\nabla_{\check{\mathsf{R}}_0}(\alpha_3) &= -\alpha_4 - \alpha_2, \\
\nabla_{\check{\mathsf{R}}_0}(\alpha_4) &= 0
\end{aligned}$$

for some regular function $\check{\mathsf{Y}}$ on T. Explicit calculations show that such a vector field and $\check{\mathsf{Y}}$ are unique [Supp Item 5]. In fact, $\check{\mathsf{Y}} = \mathsf{Y} = \frac{5^8(t_4-t_0^5)^2}{t_5^3}$ is the Yukawa coupling and

$$\check{\mathsf{R}}_0 = \mathsf{R}_0 + \frac{625(t_4 - t_0^5)}{t_5}\frac{\partial}{\partial t_6}$$

and so $\nabla_{\check{\mathsf{R}}_0}$ written in the basis $\alpha_i,\ i = 1,2,3,4$ is of the form

$$\nabla_{\check{\mathsf{R}}_0} = \begin{pmatrix} 0 & 1 & 0 & 0 \\ 0 & 0 & \frac{5^8(t_4-t_0^5)^2}{t_5^3} & 0 \\ 0 & -1 & 0 & -1 \\ 0 & 0 & 0 & 0 \end{pmatrix}. \tag{3.15}$$

Note that both $\mathsf{R} = \mathsf{R}_0, \check{\mathsf{R}}_0$ satisfy

$$\langle \alpha_1, \nabla^3_{\mathsf{R}} \alpha_1 \rangle = -\mathsf{Y}, \ \ i = 0, 1.$$

The main reason why we have taken this format of the matrix for $\nabla_{\check{\mathsf{R}}_0}$ and in particular why its $(3,2)$-entry is -1, will be explained in §4.12.

3.12 Weights

In $\mathscr{O}_{\mathsf{T}}$ we consider the weights

$$\deg(t_i) = k(i+1), \ i = 0,1,2,3,4, \ \deg(t_5) = 4k-1, \ \deg(t_6) = 3k-1$$

where k is an arbitrary natural number. In this way in the right hand side of R_0 we have homogeneous rational functions of degree $k+1, 2k+1, 3k+1, 4k+1, 5k+1, 4k, 3k$ which is compatible with the left hand side if we assume that the derivation increases the degree by one. If we would like to have the same property for $\check{\mathsf{R}}_0$ then because of the term $\frac{625(t_4-t_0^5)}{t_5}$ in $\dot{t}_6 = \cdots$, we must have the equality $5k-(4k-1) = 3k-1+1$ which implies that $k = \frac{1}{2}$. All these imply that

Proposition 6 *Let a be a non-zero constant. If t_i's are solutions of the vector field R_0 (respectively $\check{\mathsf{R}}_0$ for $k = \frac{1}{2}$) then $a^{\deg(t_i)} \cdot t_i$'s are solutions of the vector field $a^{-1} \cdot \mathsf{R}_0$ (resp. $a^{-1} \cdot \check{\mathsf{R}}_0$). Further, t_i, $i = 0,1,2,3,4, at_5, at_6$ is a solution of $a \cdot \mathsf{R}_0$.*

In Theorem 4 we have seen the first part of this proposition for $k = 3$ and $a = (\frac{2\pi i}{5})^{-1}$. The functions $(\frac{2\pi i}{5})^{-\deg(t_i)} t_i$ have q-expansion around $z = 0$ with integer coefficients. Note that if τ is the variable of derivation in R_0 and $q = e^{2\pi i \tau}$ we have $2\pi i q \frac{\partial}{\partial q} = \frac{\partial}{\partial \tau}$. Now, in $\mathscr{O}_{\mathsf{T}}$ we consider the weights

$$\hat{\deg}(t_i) = k(i+1), \ i = 0,1,2,3,4, \ \hat{\deg}(t_5) = 4k, \ \hat{\deg}(t_6) = 3k.$$

The above weights are also compatible with R_0 if we assume that the derivation does not change the degree. In other words

Proposition 7 *If t_i's are solutions of the vector field R_0 then for any non-zero constant a, $a^{\hat{\deg}(t_i)} \cdot t_i$'s are solutions of the vector field R_0.*

This also follows from a combination of the first and second part of Proposition 6.

3.13 A Lie algebra

The Lie algebra of G will play an important role in our discussion of the BCOV anomaly equation. It is given by

$$\mathrm{Lie}(\mathsf{G}) := \left\{ \mathsf{g} = [\mathsf{g}_{ij}]_{4\times 4} \in \mathrm{Mat}(4,\mathsf{k}) \mid \mathsf{g}_{ij} = 0, \text{ for } j < i \text{ and } \mathsf{g}^{\mathsf{tr}}\Phi + \Phi\mathsf{g} = 0 \right\}. \tag{3.16}$$

A basis of $\mathrm{Lie}(\mathsf{G})$ is given by

$$\mathfrak{g}_1 := \begin{pmatrix} 0 & 0 & 0 & 0 \\ 0 & -1 & 0 & 0 \\ 0 & 0 & 1 & 0 \\ 0 & 0 & 0 & 0 \end{pmatrix}, \quad \mathfrak{g}_2 := \begin{pmatrix} -1 & 0 & 0 & 0 \\ 0 & 0 & 0 & 0 \\ 0 & 0 & 0 & 0 \\ 0 & 0 & 0 & 1 \end{pmatrix}, \quad \mathfrak{g}_3 := \begin{pmatrix} 0 & 0 & 0 & 0 \\ -1 & 0 & 0 & 0 \\ 0 & 0 & 0 & 0 \\ 0 & 0 & 1 & 0 \end{pmatrix},$$

$$\mathfrak{g}_4 := \begin{pmatrix} 0 & 0 & 0 & 0 \\ 0 & 0 & 1 & 0 \\ 0 & 0 & 0 & 0 \\ 0 & 0 & 0 & 0 \end{pmatrix}, \quad \mathfrak{g}_5 := \begin{pmatrix} 0 & 0 & 1 & 0 \\ 0 & 0 & 0 & 1 \\ 0 & 0 & 0 & 0 \\ 0 & 0 & 0 & 0 \end{pmatrix}, \quad \mathfrak{g}_6 = \begin{pmatrix} 0 & 0 & 0 & 1 \\ 0 & 0 & 0 & 0 \\ 0 & 0 & 0 & 0 \\ 0 & 0 & 0 & 0 \end{pmatrix}.$$

The translation from our $\mathfrak{g}_i$ notations to those in [AMSY14] is given by

$$\mathfrak{g}_1 = \mathfrak{g}_1^1, \quad \mathfrak{g}_2 = \mathfrak{g}_0, \quad \mathfrak{g}_3 = \mathfrak{k}_1, \quad \mathfrak{g}_4 = \mathfrak{t}_{11}, \quad \mathfrak{g}_5 = \mathfrak{t}_1, \quad \mathfrak{g}_6 = \mathfrak{t}.$$

Chapter 4
Topology and periods

This chapter is dedicated to transcendental aspects of mirror quintics. By this we mean the periods of meromorphic differential 3-forms over topological cycles. We first work with periods without calculating them explicitly. We use freely the notations of Chapter 3 for $\mathsf{k} = \mathbb{C}$.

4.1 Period map

We choose a symplectic basis for $H_3(X,\mathbb{Z})$, that is, a basis $\delta_i,\ i = 1,2,3,4$ such that $[\langle \delta_i, \delta_j \rangle]$ is given by the matrix Ψ in (0.1). It is also convenient to use the basis

$$[\tilde{\delta}_1, \tilde{\delta}_2, \tilde{\delta}_3, \tilde{\delta}_4] = [\delta_1, \delta_2, \delta_3, \delta_4]\Psi^{-1} = [\delta_3, \delta_4, -\delta_1, -\delta_2].$$

In this basis the intersection form is $[\langle \tilde{\delta}_i, \tilde{\delta}_j \rangle] = \Psi^{-\mathsf{tr}} = \Psi$. Recall that in §3.6 a mirror quintic Calabi-Yau threefold X is equipped with a basis $\alpha_1, \alpha_2, \alpha_3, \alpha_4$ of $H^3_{\mathrm{dR}}(X)$ compatible with the Hodge filtration and such that $[\langle \alpha_i, \alpha_j \rangle] = \Phi$. We define the period matrix to be

$$\mathsf{P} := [\mathsf{P}_{ij}] = [\int_{\delta_i} \alpha_j]_{4\times 4}.$$

Let $\tilde{\delta}_i^{\mathsf{pd}} \in H^3(X,\mathbb{Z})$ be the Poincaré dual of $\tilde{\delta}_i$, that is, it is defined by the property

$$\int_{\delta} \tilde{\delta}_i^{\mathsf{pd}} = \langle \tilde{\delta}_i, \delta \rangle,\ \forall \delta \in H_3(\mathsf{X},\mathbb{Z}).$$

If we write α_i in terms of $\tilde{\delta}_i^{\mathsf{pd}}$ what we get is

$$[\alpha_1, \alpha_2, \alpha_3, \alpha_4] = [\tilde{\delta}_1^{\mathsf{pd}}, \tilde{\delta}_2^{\mathsf{pd}}, \tilde{\delta}_3^{\mathsf{pd}}, \tilde{\delta}_4^{\mathsf{pd}}][\int_{\delta_i} \alpha_j],$$

that is, the coefficients of the base change matrix are the periods of α_i's over δ_i's and not $\tilde{\delta}_i$'s. We have

$$[\langle \alpha_i, \alpha_j \rangle] = [\int_{\delta_i} \alpha_j]^{\mathrm{tr}} \Psi^{-\mathrm{tr}} [\int_{\delta_i} \alpha_j], \tag{4.1}$$

and so we get

$$\Phi - [\mathsf{P}_{ij}]^{\mathrm{tr}} \Psi [\mathsf{P}_{ij}] = 0. \tag{4.2}$$

This gives us 6 non-trivial polynomial relations between periods P_{ij}:

$$\begin{aligned} \mathsf{P}_{12}\mathsf{P}_{31} - \mathsf{P}_{11}\mathsf{P}_{32} + \mathsf{P}_{22}\mathsf{P}_{41} - \mathsf{P}_{21}\mathsf{P}_{42} &= 0, \\ \mathsf{P}_{13}\mathsf{P}_{31} - \mathsf{P}_{11}\mathsf{P}_{33} + \mathsf{P}_{23}\mathsf{P}_{41} - \mathsf{P}_{21}\mathsf{P}_{43} &= 0, \\ \mathsf{P}_{14}\mathsf{P}_{31} - \mathsf{P}_{11}\mathsf{P}_{34} + \mathsf{P}_{24}\mathsf{P}_{41} - \mathsf{P}_{21}\mathsf{P}_{44} + 1 &= 0, \\ \mathsf{P}_{13}\mathsf{P}_{32} - \mathsf{P}_{12}\mathsf{P}_{33} + \mathsf{P}_{23}\mathsf{P}_{42} - \mathsf{P}_{22}\mathsf{P}_{43} + 1 &= 0, \\ \mathsf{P}_{14}\mathsf{P}_{32} - \mathsf{P}_{12}\mathsf{P}_{34} + \mathsf{P}_{24}\mathsf{P}_{42} - \mathsf{P}_{22}\mathsf{P}_{44} &= 0, \\ \mathsf{P}_{14}\mathsf{P}_{33} - \mathsf{P}_{13}\mathsf{P}_{34} + \mathsf{P}_{24}\mathsf{P}_{43} - \mathsf{P}_{23}\mathsf{P}_{44} &= 0. \end{aligned} \tag{4.3}$$

These equalities correspond to the entries $(1,2), (1,3), (1,4), (2,3), (2,4)$ and $(3,4)$ of (4.2). Taking the determinant of (4.2) we see that up to sign we have $\det(\mathsf{P}) = -1$. There is another effective way to calculate this determinant without the sign ambiguity.

Proposition 8 *We have* $\det(\mathsf{P}) = -1$.

Proof. In the ideal of $\mathbb{Q}[\mathsf{P}_{ij},\ i,j = 1,2,3,4]$ generated by the polynomials

$$f_{12}, f_{13}, f_{14}, f_{23}, f_{24}, f_{34}$$

in the left hand side of (4.3), the polynomial $\det([\mathsf{P}_{ij}])$ is reduced to -1. This can be done by any software in commutative algebra which uses the Gröbner basis algorithm.

4.2 τ-locus

Let

$$C^{\mathrm{tr}} := [0,1,0,0][\langle \tilde{\delta}_i, \tilde{\delta}_j \rangle]^{-\mathrm{tr}} = [0,0,0,1].$$

We are interested in the loci L of parameters $t \in \mathsf{T}(\mathbb{C})$ such that

$$[\int_{\delta_1} \alpha_4, \ldots, \int_{\delta_4} \alpha_4] = C^{\mathrm{tr}}. \tag{4.4}$$

Using the equality corresponding to the $(1,4)$ entries of (4.1), we note that on this locus we have

$$\int_{\delta_2}\alpha_1=1,\ \int_{\delta_2}\alpha_i=0,\ i\geq 2.$$

The equalities (4.4) define a three-dimensional locus of T. We also put the following two conditions

$$\int_{\delta_1}\alpha_2=1,\ \int_{\delta_1}\alpha_3=0$$

in order to get a one-dimension locus. Finally using (4.2) we conclude that the period matrix for points in L is of the form

$$\tau=\begin{pmatrix}\tau_0 & 1 & 0 & 0\\ 1 & 0 & 0 & 0\\ \tau_1 & \tau_3 & 1 & 0\\ \tau_2 & -\tau_0\tau_3+\tau_1 & -\tau_0 & 1\end{pmatrix}. \tag{4.5}$$

The particular expressions for the $(4,2)$ and $(4,3)$ entries of the above matrix follow from the polynomial relations (4.2). The Gauss-Manin connection matrix restricted to L is

$$\mathsf{A}\mid_L=d\tau^{\mathsf{tr}}\cdot\tau^{-\mathsf{tr}}=\begin{pmatrix}0 & d\tau_0 & -\tau_3 d\tau_0+d\tau_1 & -\tau_1 d\tau_0+\tau_0 d\tau_1+d\tau_2\\ 0 & 0 & d\tau_3 & -\tau_3 d\tau_0+d\tau_1\\ 0 & 0 & 0 & -d\tau_0\\ 0 & 0 & 0 & 0\end{pmatrix}.$$

The Griffiths transversality theorem implies that

$$-\tau_3 d\tau_0+d\tau_1=0,\quad -\tau_1 d\tau_0+\tau_0 d\tau_1+d\tau_2=0.$$

Since L is one-dimensional, there are analytic relations between $\tau_i,\ i=0,1,2,3$. Therefore, we consider τ_0 as an independent parameter and τ_1,τ_2,τ_3 depending on τ_0, then we have

$$\tau_3=\frac{\partial\tau_1}{\partial\tau_0},\ \frac{\partial\tau_2}{\partial\tau_0}=\tau_1-\tau_0\frac{\partial\tau_1}{\partial\tau_0}. \tag{4.6}$$

We conclude that the Gauss-Manin connection matrix restricted to L and composed with the vector field $\frac{\partial}{\partial\tau_0}$ is given by

$$\begin{pmatrix}0 & 1 & 0 & 0\\ 0 & 0 & \frac{\partial\tau_3}{\partial\tau_0} & 0\\ 0 & 0 & 0 & -1\\ 0 & 0 & 0 & 0\end{pmatrix}. \tag{4.7}$$

Proposition 9 *The functions $t_i(\tau_0),\ i=0,1,2,\ldots,6$ obtained by the regular functions $t_i\in\mathscr{O}_{\mathsf{T}}$ restricted to L and seen as functions in τ_0, form a solution of the ordinary differential equation* R_0.

Proof. It follows from (4.7) and the uniqueness of the vector field R_0 satisfying the equalities (3.13).

4.3 Positivity conditions

The functions τ_i are originally part of a polarized Hodge structure and so they satisfy some inequalities due to the positivity condition in the definition of a polarized Hodge structure. In this section we explain this. In the case of elliptic curves we have only the quantity τ and such a positivity condition in this case implies that $\text{Im}(\tau) > 0$. For the computer codes used in the present section see the author's webpage [Supp Item 6].

Let us define the 4×4 matrix B through the equality $\alpha = B^{\text{tr}}\overline{\alpha}$. Integrating this over δ we get $\mathsf{P} = \overline{\mathsf{P}}\cdot B$ and so $B = \overline{\mathsf{P}}^{-1}\cdot\mathsf{P}$. For $1\leq i\leq 4$ let B^i be the down-left $i\times i$ sub matrix of B, i.e.

$$B = \begin{pmatrix} * & * \\ B^i & * \end{pmatrix}.$$

The pieces of the Hodge decomposition $H^3_{\text{dR}}(X) = H^{30}\oplus H^{21}\oplus H^{12}\oplus H^{03}$ are recovered from the Hodge filtration using the equalities:

$$H^{3,0} = F^3,\ H^{2,1} = F^2\cap\bar{F}^1,\ H^{12} = \overline{H^{21}},\ H^{03} = \overline{H^{30}}.$$

The fact that we get a direct sum of $H^3_{\text{dR}}(X)$ is equivalent to

$$\det(B^i)\neq 0,\ i = 1,2,3,4.$$

Using Proposition 8 we have $B = -\overline{\mathsf{P}}^{\text{adj}}\cdot\mathsf{P}$. In this way $\det(B^i)$ is a polynomial in the entries of P and $\bar{\mathsf{P}}$. Modulo the polynomial relations (4.3) we have

$$\det(B^1) = \det(B^3) = B_{4,1} = -2\sqrt{-1}\,\text{Im}(\mathsf{P}_{11}\overline{\mathsf{P}_{31}}+\mathsf{P}_{21}\overline{\mathsf{P}_{41}})$$

and $\det(B^2) = 2\text{Re}(P)$, where P is

$$\begin{aligned} P := & -\mathsf{P}_{32}\mathsf{P}_{41}\overline{\mathsf{P}}_{12}\overline{\mathsf{P}}_{21}+\mathsf{P}_{31}\mathsf{P}_{42}\overline{\mathsf{P}}_{12}\overline{\mathsf{P}}_{21}+\mathsf{P}_{32}\mathsf{P}_{41}\overline{\mathsf{P}}_{11}\overline{\mathsf{P}}_{22}-\mathsf{P}_{31}\mathsf{P}_{42}\overline{\mathsf{P}}_{11}\overline{\mathsf{P}}_{22} \quad (4.8)\\ & -\mathsf{P}_{12}\mathsf{P}_{41}\overline{\mathsf{P}}_{22}\overline{\mathsf{P}}_{31}+\mathsf{P}_{11}\mathsf{P}_{42}\overline{\mathsf{P}}_{22}\overline{\mathsf{P}}_{31}+\mathsf{P}_{12}\mathsf{P}_{41}\overline{\mathsf{P}}_{21}\overline{\mathsf{P}}_{32}-\mathsf{P}_{11}\mathsf{P}_{42}\overline{\mathsf{P}}_{21}\overline{\mathsf{P}}_{32}\\ & +2\mathsf{P}_{21}\mathsf{P}_{42}\overline{\mathsf{P}}_{22}\overline{\mathsf{P}}_{41}-|\mathsf{P}_{22}\mathsf{P}_{41}|^2-|\mathsf{P}_{21}\mathsf{P}_{42}|^2\\ = & (\mathsf{P}_{31}\mathsf{P}_{42}-\mathsf{P}_{32}\mathsf{P}_{41})(\overline{\mathsf{P}_{12}\mathsf{P}_{21}-\mathsf{P}_{11}\mathsf{P}_{22}})\\ & +(\mathsf{P}_{11}\mathsf{P}_{42}-\mathsf{P}_{12}\mathsf{P}_{41})(\overline{\mathsf{P}_{22}\mathsf{P}_{31}-\mathsf{P}_{21}\mathsf{P}_{32}})-|\mathsf{P}_{22}\mathsf{P}_{41}-\overline{\mathsf{P}_{21}\mathsf{P}_{42}}|^2. \end{aligned}$$

The basis α is compatible with the Hodge filtration. Now, we are in a position to obtain a basis compatible with the Hodge decomposition. The differential form α_1 is a basis of H^{30} and a basis of $H^{2,1}$ is given by

$$\hat{\alpha}_2 := B_{4,2}\alpha_1 - B_{4,1}\alpha_2.$$

The positivity conditions in the definition of a polarized Hodge structure are

$$(-1)^{\frac{1}{2}}\langle\alpha_1,\overline{\alpha}_1\rangle > 0, \quad (4.9)$$

$$(-1)^{\frac{3}{2}}\langle\hat{\alpha}_2,\overline{\hat{\alpha}_2}\rangle > 0, \tag{4.10}$$

We have $[\langle\alpha_i,\bar{\alpha}_j\rangle] = \mathsf{P}^{\mathsf{tr}}\Psi_0^{-\mathsf{tr}}\overline{\mathsf{P}}$ and

$$\langle\alpha_1,\bar{\alpha}_1\rangle = -\det(B^1),\ \langle\hat{\alpha}_2,\overline{\hat{\alpha}_2}\rangle = -\det(B^1)\det(B^2).$$

Therefore, the inequalities (4.9) and (4.10) are equivalent to

$$\mathrm{Im}(\mathsf{P}_{11}\overline{\mathsf{P}_{31}}+\mathsf{P}_{21}\overline{\mathsf{P}_{41}}) < 0, \tag{4.11}$$

$$\mathrm{Re}(P) < 0, \tag{4.12}$$

where P is given in (4.8). We choose the τ-locus in §4.2 and in this case the positivity conditions (4.11) and (4.12) are given by

$$\mathrm{Im}(\tau_0\bar{\tau}_1+\bar{\tau}_2) < 0, \tag{4.13}$$

$$\mathrm{Re}\,(\tau_1(-\tau_0\tau_3+\tau_1)-\tau_2\tau_3-(\tau_0(-\tau_0\tau_3+\tau_1)-\tau_2)\bar{\tau}_3) \tag{4.14}$$

$$-|-\tau_0\tau_3+\tau_1|^2 < 0.$$

4.4 Generalized period domain

We have defined the period matrix and have determined some of its properties. This leads us to the notion of (generalized) period domain in the case of mirror quintic:

$$\Pi := \left\{[\mathsf{P}_{ij}] \in \mathrm{Mat}(4\times 4,\mathbb{C}) \mid \mathsf{P}\,\text{satisfies}(4.3),(4.11),(4.12)\right\}. \tag{4.15}$$

We also define

$$\mathsf{U} := \mathrm{Sp}(4,\mathbb{Z})\backslash\Pi. \tag{4.16}$$

By definition we have a well-defined period map

$$\mathsf{P}:\mathsf{T}\to\mathsf{U}. \tag{4.17}$$

The reader may have noticed that we use the same notation P, for period map and period matrix. Hopefully, this will not produce any confusion. The Borel group G acts on P from the right by usual multiplication of matrices and the quotient

$$D := \mathsf{P}/\mathsf{G} \tag{4.18}$$

is called the Griffiths period domain. It is parametrized by $\tau_0,\tau_1,\tau_3,\tau_4$ satisfying the positivity conditions (4.13) and (4.14).

4.5 The algebraic group and τ-locus

For any 4×4 matrix P in the period domain Π satisfying

$$\mathsf{P}_{11}\mathsf{P}_{22}-\mathsf{P}_{12}\mathsf{P}_{21}\neq 0,\ \mathsf{P}_{21}\neq 0, \tag{4.19}$$

there is a unique $g\in\mathsf{G}$ such $\mathsf{P}g$ is of the form (4.5). To prove this affirmation explicitly, we take an arbitrary P and g and we write down the corresponding equations corresponding to the six entries $(2,1),(1,2),(2,2),(1,3),(2,3),(2,4)$ of $\mathsf{P}g$, that is

$$\mathsf{P}g=\begin{pmatrix} * & 1 & 0 & * \\ 1 & 0 & 0 & 0 \\ * & * & * & * \\ * & * & * & * \end{pmatrix}.$$

For our calculations we will need the coordinates of g^{-1} in terms of P_{ij}:

$$\begin{aligned}
g_1 &= \mathsf{P}_{21}^{-1},\\
g_2 &= \frac{-\mathsf{P}_{21}}{\mathsf{P}_{11}\mathsf{P}_{22}-\mathsf{P}_{12}\mathsf{P}_{21}},\\
g_3 &= \frac{-\mathsf{P}_{22}}{\mathsf{P}_{21}},\\
g_4 &= \frac{-\mathsf{P}_{12}\mathsf{P}_{23}+\mathsf{P}_{13}\mathsf{P}_{22}}{\mathsf{P}_{11}\mathsf{P}_{22}-\mathsf{P}_{12}\mathsf{P}_{21}},\\
g_5 &= \frac{\mathsf{P}_{11}\mathsf{P}_{22}\mathsf{P}_{24}-\mathsf{P}_{12}\mathsf{P}_{21}\mathsf{P}_{24}+\mathsf{P}_{12}\mathsf{P}_{22}\mathsf{P}_{23}-\mathsf{P}_{13}\mathsf{P}_{22}^2}{\mathsf{P}_{11}\mathsf{P}_{21}\mathsf{P}_{22}-\mathsf{P}_{12}\mathsf{P}_{21}^2},\\
g_6 &= \frac{\mathsf{P}_{11}\mathsf{P}_{23}-\mathsf{P}_{13}\mathsf{P}_{21}}{\mathsf{P}_{11}\mathsf{P}_{22}-\mathsf{P}_{12}\mathsf{P}_{21}}.
\end{aligned}$$

Substituting the expression of g in terms of P_{ij} in $\tau=\mathsf{P}g$ we get

$$\tau=\begin{pmatrix} \frac{\mathsf{P}_{11}}{\mathsf{P}_{21}} & 1 & 0 & 0 \\ 1 & 0 & 0 & 0 \\ \frac{\mathsf{P}_{31}}{\mathsf{P}_{21}} & \frac{-\mathsf{P}_{21}\mathsf{P}_{32}+\mathsf{P}_{22}\mathsf{P}_{31}}{\mathsf{P}_{11}\mathsf{P}_{22}-\mathsf{P}_{12}\mathsf{P}_{21}} & 1 & 0 \\ \frac{\mathsf{P}_{41}}{\mathsf{P}_{21}} & \frac{-\mathsf{P}_{21}\mathsf{P}_{42}+\mathsf{P}_{22}\mathsf{P}_{41}}{\mathsf{P}_{11}\mathsf{P}_{22}-\mathsf{P}_{12}\mathsf{P}_{21}} & -\frac{\mathsf{P}_{11}}{\mathsf{P}_{21}} & 1 \end{pmatrix}.$$

Note that for the entries $(1,4)$, $(3,3)$ and $(4,3)$ of the above matrix we have used the polynomial relations (4.2) between periods.

4.6 Monodromy covering

In §4.2 we described a solution of R_0 locally. In this section we study further such a solution in a global context. More precisely, we describe a meromorphic map $t : \mathbb{H} \to \mathsf{T}$ whose image is L defined in §4.2, where $\mathbb{H}$ is the monodromy covering of (2.25).

Let $\tilde{\mathbb{H}}$ be the moduli of the pairs (X, δ), where X is a mirror quintic Calabi-Yau threefold and $\delta = \{\delta_1, \delta_2, \delta_3, \delta_4\}$ is a basis of $H_3(\mathsf{X}, \mathbb{Z})$ such that the intersection matrix in this basis is Ψ, that is, $[\langle \delta_i, \delta_j \rangle] = \Psi$. The set $\tilde{\mathbb{H}}$ has a canonical structure of a Riemann surface, not necessarily connected. We denote by $\mathbb{H}$ the connected component of $\tilde{\mathbb{H}}$ which contains the particular pair $(\mathsf{X}_{1,z}, \delta)$ such that the monodromies around $z = 0$ and $z = 1$ are respectively given by the matrices M_0 and M_1 in the Introduction. It is already well-known that in the monodromy group $\Gamma := \langle M_0, M_1 \rangle$ the only relation between M_0 and M_1 is $(M_0 M_1)^5 = I$, see [BT14]. This is equivalent to say that $\mathbb{H}$ is biholomorphic to the upper half plane. In the following we do not need this. By definition, the monodromy group Γ acts on $\mathbb{H}$ by base change in δ. The bigger group $\mathrm{Sp}(4, \mathbb{Z})$ acts also on $\tilde{\mathbb{H}}$ by base change and all connected components of $\tilde{\mathbb{H}}$ are obtained by $\mathbb{H}_\alpha := \alpha(\mathbb{H}),\ \alpha \in \mathrm{Sp}(4, \mathbb{Z})/\Gamma$:

$$\tilde{\mathbb{H}} := \bigcup_{\alpha \in \mathrm{Sp}(4,\mathbb{Z})/\Gamma} \tilde{\mathbb{H}}_\alpha.$$

From now on by w we denote a point (X, δ) of $\mathbb{H}$. We use the following meromorphic functions on $\mathbb{H}$:

$$\tau_i : \mathbb{H} \to \mathbb{C},\ i = 0, 1, 2,$$

$$\tau_0(w) = \frac{\int_{\delta_1} \alpha_1}{\int_{\delta_2} \alpha_1},\quad \tau_1(w) = \frac{\int_{\delta_3} \alpha_1}{\int_{\delta_2} \alpha_1},\quad \tau_2(w) = \frac{\int_{\delta_4} \alpha_1}{\int_{\delta_2} \alpha_1},$$

where α_1 is a holomorphic differential form on X. They do not depend on the choice of α_1. For simplicity, we have used the same notations τ_i as in §4.

There is a useful meromorphic function z on $\mathbb{H}$ which is obtained by identifying X with some $\mathsf{X}_{1,z}$. It has a pole of order 5 at elliptic points which are the pairs (X, δ) with $X = \mathsf{X}_{\psi,1},\ \psi = 0$. In this way, we have a well-defined holomorphic function

$$\psi = z^{-\frac{1}{5}} : \mathbb{H} \to \mathbb{C}.$$

The coordinate system τ_0 is adapted for calculations around the cusp $z = 0$. Let B be the set of points $w = (X, \delta) \in \mathbb{H}$ such that either τ_0 has a pole at w or it has a critical point at w, that is, $\frac{\partial \tau_0}{\partial z}(w) = 0$. We do not know whether B is empty or not. Many functions that we are going to study are meromorphic with poles at B. The set B is characterized by this property that in its complement in $\mathbb{H}$ the inequalities (4.19) hold.

4.7 A particular solution

For a point $w = (X, \delta) \in \mathbb{H} \backslash B$ there is a unique basis α of $H^3_{\mathrm{dR}}(X)$ such that (X, α) is an element in the moduli space $\mathsf{T}(\mathbb{C})$ defined in §3.6 and the period matrix $[\int_{\delta_i} \alpha_j]$ of the triple (X, δ, α) is of the form (4.5). This follows from the arguments in §4.5. In this way we have well-defined meromorphic maps

$$t : \mathbb{H} \to \mathsf{T}$$

and

$$\tau : \mathbb{H} \to \Pi$$

which are characterized by the uniqueness of the basis α and the equality

$$\tau(w) = [\int_{\delta_i} \alpha_j].$$

If we parameterize $\mathbb{H}$ by the image of τ_0 then t is the same map as in §4.2. We conclude that the map $t : \mathbb{H} \to \mathsf{T}$ with the coordinate system τ_0 on $\mathbb{H}$ is a solution of R_0. The functions t and τ are holomorphic outside the poles and critical points of τ_0 (this corresponds to points in which the inequalities (4.19) occur).

4.8 Action of the monodromy

The monodromy group $\Gamma := \langle M_0, M_1 \rangle$ acts on $\mathbb{H}$ by base change. If we choose the local coordinate system τ_0 on $\mathbb{H}$ then this action is given by

$$A(\tau_0) = \frac{a_{11}\tau_0 + a_{12} + a_{13}\tau_1 + a_{14}\tau_2}{a_{21}\tau_0 + a_{22} + a_{23}\tau_1 + a_{24}\tau_2}, \quad A = [a_{ij}] \in \Gamma.$$

Proposition 10 *For all $A \in \Gamma$ we have*

$$t(w) = t(A(w)) \bullet \mathsf{g}(A, w),$$

where $\mathsf{g}(A, w) \in \mathsf{G}$ is defined using the equality

$$A \cdot \tau(w) = \tau(A(w)) \cdot \mathsf{g}(A, w).$$

Proof. Let $w = (X, \delta) \in \mathbb{H}$ and $t(w) = (X, \alpha)$. By definition we have

$$[\int_{A(\delta)_i} \alpha_j]\mathsf{g}(A, w)^{-1} = A\tau(w)\mathsf{g}(A, w)^{-1} = \tau(A(w)).$$

Therefore, $t(A(w)) = (X, \alpha \cdot \mathsf{g}(A, w)^{-1}) = t(w) \bullet \mathsf{g}(A, w)^{-1}$.

If we choose the coordinate system τ_0 on $\mathbb{H}$ and regard the parameters t_i's and τ_i's as functions in τ_0, then we have

$$t(\tau_0) = t(A(\tau_0)) \bullet \mathsf{g}(A, \tau_0).$$

These are the functional equations of $t_i(\tau_0)$'s mentioned in the Introduction. For $A = M_0$ we have

$$\begin{pmatrix} 1 & 1 & 0 & 0 \\ 0 & 1 & 0 & 0 \\ 5 & 5 & 1 & 0 \\ 0 & -5 & -1 & 1 \end{pmatrix} \begin{pmatrix} \tau_0 & 1 & 0 & 0 \\ 1 & 0 & 0 & 0 \\ \tau_1 & \tau_3 & 1 & 0 \\ \tau_2 & -\tau_0\tau_3+\tau_1 & -\tau_0 & 1 \end{pmatrix} = \begin{pmatrix} \tau_0+1 & 1 & 0 & 0 \\ 1 & 0 & 0 & 0 \\ \tau_1+5\tau+5 & \tau_3+5 & 1 & 0 \\ \tau_2-5-\tau_1 & -\tau_0(\tau_3+1)+\tau_1 & -\tau_0-1 & 1 \end{pmatrix}$$

which is already of the format (4.5). Note that

$$-(\tau_0+1)(\tau_3+5)+\tau_1+5\tau_0+5 = -\tau_0(\tau_3+1)+\tau_1.$$

Therefore, $M_0(\tau_0) = \tau_0+1$ and $\mathsf{g}(M_0, \tau_0)$ is the identity matrix. The corresponding functional equation of t_i simply says that t_i is invariant under $\tau_0 \mapsto \tau_0+1$:

$$t_i(\tau_0) = t_i(\tau_0+1), \ i = 0, 1, \ldots, 6.$$

For $A = M_1$ we have

$$\begin{pmatrix} 1 & 0 & 0 & 0 \\ 0 & 1 & 0 & 1 \\ 0 & 0 & 1 & 0 \\ 0 & 0 & 0 & 1 \end{pmatrix} \begin{pmatrix} \tau_0 & 1 & 0 & 0 \\ 1 & 0 & 0 & 0 \\ \tau_1 & \tau_3 & 1 & 0 \\ \tau_2 & -\tau_0\tau_3+\tau_1 & -\tau_0 & 1 \end{pmatrix}$$

$$= \begin{pmatrix} \tau_0 & 1 & 0 & 0 \\ \tau_2+1 & -\tau_0\tau_3+\tau_1 & -\tau_0 & 1 \\ \tau_1 & \tau_3 & 1 & 0 \\ \tau_2 & -\tau_0\tau_3+\tau_1 & -\tau_0 & 1 \end{pmatrix}$$

$$= \begin{pmatrix} \frac{\tau_0}{\tau_2+1} & 1 & 0 & 0 \\ 1 & 0 & 0 & 0 \\ \frac{\tau_1}{\tau_2+1} & \frac{\tau_0\tau_1\tau_3-\tau_1^2+\tau_2\tau_3+\tau_3}{\tau_0^2\tau_3-\tau_0\tau_1+\tau_2+1} & 1 & 0 \\ \frac{\tau_2}{\tau_2+1} & \frac{-\tau_0\tau_3+\tau_1}{\tau_0^2\tau_3-\tau_0\tau_1+\tau_2+1} & \frac{-\tau_0}{\tau_2+1} & 1 \end{pmatrix}$$

$$\cdot \begin{pmatrix} (\tau_2+1) & (-\tau_0\tau_3+\tau_1) & (-\tau_0) & 1 \\ 0 & \frac{\tau_0^2\tau_3-\tau_0\tau_1+\tau_2+1}{\tau_2+1} & \frac{\tau_0^2}{\tau_2+1} & \frac{-\tau_0}{\tau_2+1} \\ 0 & 0 & \frac{\tau_2+1}{\tau_0^2\tau_3-\tau_0\tau_1+\tau_2+1} & \frac{\tau_0\tau_3-\tau_1}{\tau_0^2\tau_3-\tau_0\tau_1+\tau_2+1} \\ 0 & 0 & 0 & \frac{1}{\tau_2+1} \end{pmatrix},$$

where the element of the algebraic group G in the right hand side has the coordinates:

$$g_1 = \frac{1}{\tau_2+1},$$
$$g_2 = \frac{\tau_2+1}{\tau_0^2\tau_3 - \tau_0\tau_1 + \tau_2 + 1},$$
$$g_3 = \frac{\tau_0\tau_3 - \tau_1}{\tau_2+1},$$
$$g_4 = \frac{-\tau_0}{\tau_0^2\tau_3 - \tau_0\tau_1 + \tau_2 + 1},$$
$$g_5 = \frac{1}{\tau_0^2\tau_3 - \tau_0\tau_1 + \tau_2 + 1},$$
$$g_6 = \frac{\tau_0^2}{\tau_0^2\tau_3 - \tau_0\tau_1 + \tau_2 + 1}.$$

In this case we have
$$M_1(\tau_0) = \frac{\tau_0}{\tau_2+1}.$$

The corresponding functional equations of t_i's can be written immediately. These are presented in Theorem 2.

4.9 The solution in terms of periods

In this section we explicitly calculate the map t. For $w = (X, \delta) \in \mathbb{H}$ we identify X with $\mathsf{X}_{1,z}$ and hence we obtain a unique point $\tilde{z} = (1,0,0,0,z,1,0) \in \mathsf{T}$. Now, we have a well-defined period map

$$\mathsf{P} : \mathbb{H} \to \Pi,$$
$$w = (\mathsf{X}_{1,z}, \delta) \mapsto [\int_{\delta_i} \alpha_j].$$

We write $\mathsf{P}(w)\mathsf{g}(w) = \tau(w)$, where $\tau(w)$ is of the form (4.5) and $\mathsf{g}(w) \in \mathsf{G}$. We have

$$t(w) = \tilde{z} \bullet \mathsf{g}(w).$$

For the one-dimensional locus $\tilde{z} \in \mathsf{T}$, we have $\alpha = S\omega$ and $\omega = T\tilde{\eta}$ [Supp Item 7], where

$$S = \begin{pmatrix} 1 & 0 & 0 & 0 \\ -5^5 & -5^4(z-1) & 0 & 0 \\ -\frac{5}{z-1} & 0 & 1 & 0 \\ 0 & 0 & 0 & 5^4(z-1) \end{pmatrix}, \quad T = \begin{pmatrix} 1 & 0 & 0 & 0 \\ -1 & -5 & 0 & 0 \\ 2 & 15 & 25 & 0 \\ -6 & -55 & -150 & -125 \end{pmatrix},$$

$$\tilde{\eta} = [\eta, \theta\eta, \theta^2\eta, \theta^3\eta]^{\mathrm{tr}}, \quad \theta = z\frac{\partial}{\partial z},$$

and ω is given in (3.5). Therefore, $\alpha = ST\eta$. Restricted to $\tilde{z}$-locus we have $\alpha_1 = \omega_1 = \eta$ and by our definition of x_{ij}'s in the introduction

$$x_{ij} = \theta^{j-1} \int_{\delta_i} \eta, \ i,j = 1,2,3,4.$$

Therefore,

$$\mathsf{P}(w) = [x_{ij}](ST)^{\mathsf{tr}}. \tag{4.20}$$

Now, the map $w \mapsto t(w)$, where the domain $\mathbb{H}$ is equipped with the coordinate system z, is given by the expressions for t_i in Theorem 4. We conclude that if we write t_i's in terms of τ_0 then we get functions which are solutions to R_0. Note that

$$\frac{\partial}{\partial \tau_0} = 2\pi i q \frac{\partial}{\partial q} = \left(z\frac{\partial \frac{x_{11}}{x_{21}}}{\partial z}\right)^{-1} z\frac{\partial}{\partial z} = \frac{x_{21}^2}{x_{12}x_{21} - x_{11}x_{22}}\theta.$$

4.10 Computing periods

In this section we calculate the periods x_{ij} explicitly. This will finish the proof of Theorems 4, 2. We restrict the parameter $t \in \mathsf{T}(\mathbb{C})$ to the one-dimensional locus $\tilde{z}$ given by $t_0 = 1, t_1 = t_2 = t_3 = 0, t_4 = z, t_5 = 1, t_6 = 0$. On this locus $\eta = \omega_1 = \alpha_1$. We know that the integrals $\int_\delta \eta,\ \delta \in H_3(\mathsf{X}_{1,z}, \mathbb{Q})$ satisfy the linear differential equation (3.9). Four linearly independent solutions of (3.9) are given by $\psi_0, \psi_1, \psi_2, \psi_3$ in the Introduction, see for instance [vEvS06] and [CYY08]. In fact, there are four topological cycles with complex coefficients $\hat{\delta}_1, \hat{\delta}_2, \hat{\delta}_3, \hat{\delta}_4 \in H_3(\mathsf{X}_{1,z}, \mathbb{C})$ such that $\int_{\hat{\delta}_i} \eta = \frac{(2\pi i)^{i-1}}{5^4} \psi_{4-i}$. Note that the pair $(\mathsf{X}_{1,z}, 5\eta)$ is isomorphic to the pair $(\mathsf{X}_\psi, \Omega)$ used in [CdlOGP91]. We use a new basis given by

$$\begin{pmatrix} \delta_1 \\ \delta_2 \\ \delta_3 \\ \delta_4 \end{pmatrix} = \begin{pmatrix} 0 & 0 & 1 & 0 \\ 0 & 0 & 0 & 1 \\ 0 & d & \frac{d}{2} & -b \\ -d & 0 & -b & -a \end{pmatrix} \begin{pmatrix} \hat{\delta}_1 \\ \hat{\delta}_2 \\ \hat{\delta}_3 \\ \hat{\delta}_4 \end{pmatrix},$$

where

$$a = \frac{c_3}{(2\pi i)^3}\zeta(3) = \frac{-200}{(2\pi i)^3}\zeta(3), \ b = c_2 \cdot H/24 = \frac{25}{12}, \ d = H^3 = 5$$

(these notations are used in [vEvS06]). The monodromies around $z = 0$ and $z = 1$ written in the basis δ_i are respectively

$$M_0 := \begin{pmatrix} 1 & 1 & 0 & 0 \\ 0 & 1 & 0 & 0 \\ d & d & 1 & 0 \\ 0 & -k & -1 & 1 \end{pmatrix} \quad M_1 := \begin{pmatrix} 1 & 0 & 0 & 0 \\ 0 & 1 & 0 & 1 \\ 0 & 0 & 1 & 0 \\ 0 & 0 & 0 & 1 \end{pmatrix},$$

where $k = 2b + \frac{d}{6} = 5$, see [CYY08]. In fact, $\delta_i \in H_3(\mathsf{X}_{1,z}, \mathbb{Z})$, $i = 1,2,3,4$. This follows from the calculations in [CdlOGP91] and the expressions for monodromy matrices. In summary, we have

$$x_{11} = \int_{\delta_1} \eta = \frac{1}{5^2}(\frac{2\pi i}{5})^2 \psi_1(\tilde{z}),$$
$$x_{21} = \int_{\delta_2} \eta = \frac{1}{5}(\frac{2\pi i}{5})^3 \psi_0,$$
$$x_{31} = \int_{\delta_3} \eta = \frac{d}{125}\psi_2(\tilde{z})\frac{2\pi i}{5} + \frac{d}{50}\cdot(\frac{2\pi i}{5})^2 \cdot \psi_1(\tilde{z}) - \frac{b}{5}\cdot(\frac{2\pi i}{5})^3 \cdot \psi_0(\tilde{z}),$$
$$x_{41} = \int_{\delta_4} \eta = \frac{-d}{5^4}\psi_3(\tilde{z}) + \frac{-b}{5^2}\cdot(\frac{2\pi i}{5})^2 \cdot \psi_1(\tilde{z}) - \frac{a}{5}\cdot(\frac{2\pi i}{5})^3 \cdot \psi_0(\tilde{z}),$$

where $\tilde{z} = \frac{z}{5^5}$. We have also

$$\tau_0 = \frac{\int_{\delta_1} \eta}{\int_{\delta_2} \eta} = \frac{1}{2\pi i}\frac{\psi_1(\tilde{z})}{\psi_0(\tilde{z})},$$
$$\tau_1 = \frac{\int_{\delta_3} \eta}{\int_{\delta_2} \eta} = d(\frac{1}{2}\tau_0^2 + \frac{1}{5}H') + \frac{d}{2}\tau_0 - b = -b + \frac{d}{2}\tau_0(\tau_0+1) + \frac{d}{5}H',$$
$$\tau_2 = \frac{\int_{\delta_4} \eta}{\int_{\delta_2} \eta} = -d(\frac{-1}{3}\tau_0^3 + \tau_0(\frac{1}{2}\tau_0^2 + \frac{1}{5}H') + \frac{2}{5}H) - b\tau_0 - a$$
$$= -a - b\tau_0 - \frac{d}{6}\tau_0^3 - \frac{d}{5}\tau_0 H' - \frac{2d}{5}H,$$

where H is defined in (2.35). We have used the equalities

$$\frac{\psi_2}{\psi_0} - \frac{1}{2}(\frac{\psi_1}{\psi_0})^2 = \frac{1}{5}(\sum_{n=1}^{\infty}(\sum_{d|n} n_d d^3)\frac{q^n}{n^2}),$$

$$\frac{1}{3}(\frac{\psi_1}{\psi_0})^3 - \frac{\psi_1}{\psi_0}\frac{\psi_2}{\psi_0} + \frac{\psi_3}{\psi_0} = \frac{2}{5}\sum_{n=1}^{\infty}(\sum_{d|n} n_d d^3)\frac{q^n}{n^3},$$

see for instance [Kon95, Pan98]. We can use the explicit series

$$\psi_0(\tilde{z}) = \sum_{m=0}^{\infty}\frac{(5m)!}{(m!)^5}\tilde{z}^m,$$

$$\psi_1(\tilde{z}) = \ln(\tilde{z})\psi_0(\tilde{z}) + 5\tilde{\psi}_1(\tilde{z}), \quad \tilde{\psi}_1(\tilde{z}) = \sum_{m=1}^{\infty}\frac{(5m)!}{(m!)^5}(\sum_{k=m+1}^{5m}\frac{1}{k})\tilde{z}^m$$

and calculate the q-expansion of $t_i(\tau_0)$ around the cusp $z = 0$. There is another way of doing this using the differential equation R_0. We just use the above equalities to obtain the initial values (2.29) in the Introduction. We write each t_i as a formal power series in q, $t_i = \sum_{n=0}^{\infty} t_{i,n}q^n$, and substitute in (2.9) with $\dot{t} := 5q\frac{\partial t}{\partial q}$. Let

$$T_n = [t_{0,n}, t_{1,n}, t_{2,n}, t_{3,n}, t_{4,n}, t_{5,n}, t_{6,n}].$$

Comparing the coefficients of q^0 and q^1 in both sides of R_0 we get

$$T_0 = [\frac{1}{5}, -25, -35, -6, 0, -1, -15],$$

$$T_1 = [24, -2250, -5350, -355, 1, 1875, 4675].$$

Comparing the coefficients of $q^n, n \geq 2$ we get a recursion of the following type:

$$(A_0 + 5nI_{7\times 7})T_n^{\mathrm{tr}} = \text{A function of the entries of } T_0,\ T_1, \ldots, T_{n-1},$$

where

$$A_0 = [\frac{\partial(t_5 \mathsf{R}_{0,i})}{\partial t_j}]_{i,j=0,1,\ldots,6} \text{ evaluated at } t = T_0,\quad \mathsf{R}_0 = \sum_{i=0}^{6} \mathsf{R}_{0,i} \frac{\partial}{\partial t_i}.$$

The matrix $A_0 + 5nI_{7\times 7},\ n \geq 2$ is invertible and so we get a recursion in T_n.

Remark 4 There is an elegant way to compute the period ψ_0. This is as follows. Let us write W_z in (3.1) in the affine coordinates given by $x_0 = 1$. We define $y_i := z^{-\frac{1}{5}} \frac{x_i^5}{x_1 x_2 x_3 x_4}$, $i = 1,2,3,4$ which are invariant under the action of G. This gives us an affine chart for X. In fact, in the affine coordinate system (y_1, y_2, y_3, y_4), we can write X in the following format

$$X:\ a \cdot f - 1 = 0, \tag{4.21}$$

$$f := y_1 + y_2 + y_3 + y_4 + \frac{1}{y_1 y_2 y_3 y_4},\quad a := \frac{1}{5} z^{\frac{1}{5}}.$$

Let $\Gamma = S^1 \times S^1 \times S^1 \times S^1 \subset \mathbb{C}^4$, where S^1 is the circle in $\mathbb{C}$ with center 0 and radius one. We have

$$\int_\Gamma \frac{1}{1 - a \cdot f} \frac{dy_1}{y_1} \wedge \frac{dy_2}{y_2} \wedge \frac{dy_3}{y_3} \wedge \frac{dy_4}{y_4} = \int_\Gamma \sum_{n=0}^{\infty} a^n f(y)^n \frac{dy_1}{y_1} \wedge \frac{dy_2}{y_2} \wedge \frac{dy_3}{y_3} \wedge \frac{dy_4}{y_4}$$
$$= (2\pi i)^4 \sum_{n=0}^{\infty} \frac{(5n)!}{(n!)^5} a^{5n}.$$

For more examples of computing periods in this style see [GAZ10] and the references therein.

4.11 Algebraically independent periods

In this section we prove Theorem 6. We plug the period expression (4.20) into (4.3) and get six polynomial relations between x_{ij} with coefficients in $\mathbb{C}(z)$. Let us call them $f_{12}, f_{13}, f_{14}, f_{23}, f_{24}, f_{34}$. Let y_{ij} be indeterminate variables, $R = \mathbb{C}(z)[y_{ij}, i, j = 1,2,3,4]$ and

$$I := \{f \in R \mid f(x_{ij}) = 0\}. \tag{4.22}$$

Proposition 11 *The ideal I is generated by $f_{12}, f_{13}, f_{14}, f_{23}, f_{24}, f_{34}$.*

Proof. Let E be the differential field over $F = \mathbb{C}(z)$ generated by x_{ij}'s. Note that the matrix $[x_{ij}]$ is the fundamental system of the linear differential equation

$$\frac{\partial}{\partial z}[x_{ij}] = [x_{ij}]B(z)^{\mathsf{tr}},$$

where $B(z)$ is obtained from the matrix (3.11) by putting $dt_0 = 0,\ dt_4 = 1,\ t_0 = 1,\ t_4 = z$. The homology group $H_3(X_{1,z}, \mathbb{Q})$ has a symplectic basis and hence the monodromy group of $X_{1,z}$ is a subgroup of $\mathrm{Sp}(4,\mathbb{Z})$. Consequently, the differential Galois group $G(E/F)$ is an algebraic subgroup of $\mathrm{Sp}(4,\mathbb{C})$ and it contains a maximal unipotent matrix which is the monodromy around $z = 0$. By a result of Saxl and Seitz, see [SS97], we have $G(E/F) = \mathrm{Sp}(4,\mathbb{C})$. Therefore, $\dim G(E/F) = 10$ which is the transcendental degree of the field E over F (see [vdPS03]).

Proof of Theorem 6: First, we note that if there is a polynomial relation with coefficients in $\mathbb{C}$ between $h_i, i = 0, 1, \ldots, 6$ (as power series in $q = e^{2\pi i \tau_0}$ and hence as functions in τ_0) then the same is true if we change the variable τ_0 by some function in another variable. In particular, we put $\tau_0 = \frac{x_{21}}{x_{11}}$ and obtain t_i's in terms of periods. Now, it is enough to prove that the period expressions in Theorem 4 are algebraically independent over $\mathbb{C}$. Using Proposition 11, it is enough to prove that the variety induced by the ideal $\tilde{I} = \langle t_i - k_i,\ i = 0, 1, \ldots, 6\rangle + I \subset k[y_{ij}, i, j = 1,2,3,4]$ is of dimension $16 - 6 - 7 = 3$. Here k_i's are arbitrary parameters, I is the ideal in (4.22), $\mathsf{k} = \mathbb{C}(k_i, i = 0, 1, \ldots, 6)$ and in the expressions of t_i we have written y_{ij} instead of x_{ij}. This can be done by any software in commutative algebra (see for instance [GPS01]).

4.12 θ-locus

In Theorem 4 the topological cycles involved in the nominator and denominator of the mirror map have zero intersection. Later, in Theorem 10 we will see that if such a number is one then the resulting period expressions are solutions of $\check{\mathsf{R}}_0$. We did not explain why we have used the matrix (3.15) in order to calculate $\check{\mathsf{R}}_0$. In this section we explain all these missing arguments [Supp Item 8].

We are interested in the subset $\check{L}$ of $\mathsf{T}(\mathbb{C})$ such that the period matrix restricted to $\check{L}$ is of the form

$$\mathsf{P} = \begin{pmatrix} * & * & * & 0 \\ * & 1 & 0 & -1 \\ * & * & * & 0 \\ * & * & * & 0 \end{pmatrix}.$$

From the data of the last column we conclude that $\mathsf{P}_{41} = 1$ and this is the main reason why we choose the last column of the above matrix. It also follows that $\mathsf{P}_{42} = \mathsf{P}_{43} = 0$, $\mathsf{P}_{13}\mathsf{P}_{32} - \mathsf{P}_{12}\mathsf{P}_{33} = -1$, $\mathsf{P}_{31} = \mathsf{P}_{11}$ and $\mathsf{P}_{33} = -\mathsf{P}_{31}$. Finally, we get the following format for the period matrix:

$$\theta = \begin{pmatrix} \theta_1 & \theta_3 & -\theta_1 & 0 \\ \theta_0 & 1 & 0 & -1 \\ \theta_2 & \frac{1+\theta_2\theta_3}{\theta_1} & -\theta_2 & 0 \\ 1 & 0 & 0 & 0 \end{pmatrix}. \tag{4.23}$$

We define θ_4 to be the $(3,2)$-entry of the above matrix and so $\theta_1\theta_4 - \theta_2\theta_3 = 1$. Now the Gauss-Manin connection matrix restricted to $\hat{L}$ is given by

$$\mathsf{A}\mid_{\hat{L}} = d\mathsf{P}^{\mathsf{tr}} \cdot \mathsf{P}^{-\mathsf{tr}}\mid_{\hat{L}}$$

$$= \begin{pmatrix} 0 & -\theta_2 d\theta_1 + \theta_1 d\theta_2 & -\theta_4 d\theta_1 + \theta_3 d\theta_2 & -d\theta_0 - \theta_2 d\theta_1 + \theta_1 d\theta_2 \\ 0 & -\theta_2 d\theta_3 + \theta_1 d\theta_4 & -\theta_4 d\theta_3 + \theta_3 d\theta_4 & -\theta_2 d\theta_3 + \theta_1 d\theta_4 \\ 0 & \theta_2 d\theta_1 - \theta_1 d\theta_2 & \theta_4 d\theta_1 - \theta_3 d\theta_2 & \theta_2 d\theta_1 - \theta_1 d\theta_2 \\ 0 & 0 & 0 & 0 \end{pmatrix}.$$

The Griffiths transversality distribution is given by $(1,3)$ $(1,4)$ and $(2,4)$ entries of the above matrix to be zero:

$$\begin{aligned} -\theta_4 d\theta_1 + \theta_3 d\theta_2 &= 0, \\ -d\theta_0 - \theta_2 d\theta_1 + \theta_1 d\theta_2 &= 0, \\ -\theta_2 d\theta_3 + \theta_1 d\theta_4 &= 0. \end{aligned}$$

Modulo these differential relations we have

$$\mathrm{GM}\mid_{\hat{L}} = \begin{pmatrix} 0 & d\theta_0 & 0 & 0 \\ 0 & 0 & -\theta_4 d\theta_3 + \theta_3 d\theta_4 & 0 \\ 0 & -d\theta_0 & 0 & -d\theta_0 \\ 0 & 0 & 0 & 0 \end{pmatrix}.$$

We consider θ_0 as an independent parameter and θ_i, $i = 1,2,3,4$ depending on θ_0. In this way the Gauss-Manin connection matrix composed with the vector field $\frac{\partial}{\partial\theta_0}$ is given by

$$\begin{pmatrix} 0 & 1 & 0 & 0 \\ 0 & 0 & -\theta_4 \frac{d\theta_3}{d\theta} + \theta_3 \frac{d\theta_4}{d\theta} & 0 \\ 0 & -1 & 0 & -1 \\ 0 & 0 & 0 & 0 \end{pmatrix}$$

and the Griffiths transversality gives us:

$$-\theta_4\frac{d\theta_1}{d\theta_0}+\theta_3\frac{d\theta_2}{d\theta_0}=0,\quad -1-\theta_2\frac{d\theta_1}{d\theta_0}+\theta_1\frac{d\theta_2}{d\theta_0}=0.$$

As in the case of L and R_0, it follows that the functions $\hat{t}_i(\theta_0)$ obtained by the regular functions t_i, $i=0,1,2,\ldots,6$ restricted to $\hat{L}$ and seen as functions in θ_0 form a solution of the ordinary differential equation $\check{\mathsf{R}}_0$.

4.13 The algebraic group and the θ-locus

In a similar way as in the case of τ-locus, we start to find an element $\mathsf{g}\in\mathsf{G}$ according to the equality

$$\mathsf{P}\mathsf{g}=\begin{pmatrix} * & * & * & * \\ * & 1 & 0 & * \\ * & * & * & * \\ 1 & 0 & 0 & 0 \end{pmatrix}.$$

We get

$$\mathsf{g}=\begin{pmatrix} \frac{1}{\mathsf{P}_{41}} & \frac{\mathsf{P}_{42}}{\mathsf{P}_{21}\mathsf{P}_{42}-\mathsf{P}_{22}\mathsf{P}_{41}} & \frac{-\mathsf{P}_{22}\mathsf{P}_{43}+\mathsf{P}_{23}\mathsf{P}_{42}}{\mathsf{P}_{41}} & -\mathsf{P}_{44} \\ 0 & \frac{-\mathsf{P}_{41}}{\mathsf{P}_{21}\mathsf{P}_{42}-\mathsf{P}_{22}\mathsf{P}_{41}} & \frac{\mathsf{P}_{21}\mathsf{P}_{43}-\mathsf{P}_{23}\mathsf{P}_{41}}{\mathsf{P}_{41}} & -\mathsf{P}_{43} \\ 0 & 0 & \frac{-\mathsf{P}_{21}\mathsf{P}_{42}+\mathsf{P}_{22}\mathsf{P}_{41}}{\mathsf{P}_{41}} & \mathsf{P}_{42} \\ 0 & 0 & 0 & \mathsf{P}_{41} \end{pmatrix}$$

with g_i's:

$$\begin{aligned}
\mathsf{g}_1 &= \mathsf{P}_{41},\\
\mathsf{g}_2 &= \frac{-\mathsf{P}_{21}\mathsf{P}_{42}+\mathsf{P}_{22}\mathsf{P}_{41}}{\mathsf{P}_{41}},\\
\mathsf{g}_3 &= \frac{-\mathsf{P}_{41}\mathsf{P}_{42}}{\mathsf{P}_{21}\mathsf{P}_{42}-\mathsf{P}_{22}\mathsf{P}_{41}},\\
\mathsf{g}_4 &= \frac{\mathsf{P}_{21}\mathsf{P}_{42}\mathsf{P}_{43}-\mathsf{P}_{22}\mathsf{P}_{41}\mathsf{P}_{43}}{\mathsf{P}_{41}},\\
\mathsf{g}_5 &= -\mathsf{P}_{41}\mathsf{P}_{44}-\mathsf{P}_{42}\mathsf{P}_{43},\\
\mathsf{g}_6 &= \frac{-\mathsf{P}_{21}^2\mathsf{P}_{42}\mathsf{P}_{43}+\mathsf{P}_{21}\mathsf{P}_{22}\mathsf{P}_{41}\mathsf{P}_{43}+\mathsf{P}_{21}\mathsf{P}_{23}\mathsf{P}_{41}\mathsf{P}_{42}-\mathsf{P}_{22}\mathsf{P}_{23}\mathsf{P}_{41}^2}{\mathsf{P}_{41}^2}
\end{aligned}$$

and

$$\theta=\begin{pmatrix} \frac{\mathsf{P}_{11}}{\mathsf{P}_{41}} & \frac{\mathsf{P}_{11}\mathsf{P}_{42}-\mathsf{P}_{12}\mathsf{P}_{41}}{\mathsf{P}_{21}\mathsf{P}_{42}-\mathsf{P}_{22}\mathsf{P}_{41}} & \frac{-\mathsf{P}_{11}}{\mathsf{P}_{41}} & 0 \\ \frac{\mathsf{P}_{21}}{\mathsf{P}_{41}} & 1 & 0 & -1 \\ \frac{\mathsf{P}_{31}}{\mathsf{P}_{41}} & \frac{\mathsf{P}_{31}\mathsf{P}_{42}-\mathsf{P}_{32}\mathsf{P}_{41}}{\mathsf{P}_{21}\mathsf{P}_{42}-\mathsf{P}_{22}\mathsf{P}_{41}} & \frac{-\mathsf{P}_{31}}{\mathsf{P}_{41}} & 0 \\ 1 & 0 & 0 & 0 \end{pmatrix}.$$

4.14 Comparing τ and θ-loci

For a matrix τ described in §4.2, there is a unique $\mathrm{g} \in \mathrm{G}$ such that $\tau \mathrm{g} = \theta$ is of the form (4.23):

$$\mathrm{g} = \begin{pmatrix} \frac{1}{\tau_2} & 1 & 0 & -1 \\ 0 & \frac{\tau_2}{\tau_0\tau_3-\tau_1} & \frac{-\tau_0}{\tau_2} & \tau_0 \\ 0 & 0 & \frac{\tau_0\tau_3-\tau_1}{\tau_2} & -\tau_0\tau_3+\tau_1 \\ 0 & 0 & 0 & \tau_2 \end{pmatrix}, \quad \theta = \begin{pmatrix} \frac{\tau_0}{\tau_2} & \frac{\tau_0^2\tau_3-\tau_0\tau_1+\tau_2}{\tau_0\tau_3-\tau_1} & \frac{-\tau_0}{\tau_2} & 0 \\ \frac{1}{\tau_2} & 1 & 0 & -1 \\ \frac{\tau_1}{\tau_2} & \frac{\tau_0\tau_1\tau_3-\tau_1^2+\tau_2\tau_3}{\tau_0\tau_3-\tau_1} & \frac{-\tau_1}{\tau_2} & 0 \\ 1 & 0 & 0 & 0 \end{pmatrix}.$$

The matrix g has the coordinates

$$\begin{aligned} \mathrm{g}_1 &= \tau_2, \\ \mathrm{g}_2 &= \frac{\tau_0\tau_3-\tau_1}{\tau_2}, \\ \mathrm{g}_3 &= -\tau_2, \\ \mathrm{g}_4 &= \frac{\tau_0^2\tau_3-\tau_0\tau_1}{\tau_2}, \\ \mathrm{g}_5 &= -\tau_0^2\tau_3+\tau_0\tau_1-\tau_2, \\ \mathrm{g}_6 &= \frac{-\tau_0^2\tau_3+\tau_0\tau_1}{\tau_2^2}. \end{aligned}$$

We conclude that

Proposition 12 *The functions $t_i(\tau_0)$ and $\hat{t}_i(\theta_0)$ are related through the functional equations:*

$$\begin{aligned} \hat{t}_0(\frac{1}{\tau_2}) &= t_0(\tau_0)\cdot\tau_2, \\ \hat{t}_1(\frac{1}{\tau_2}) &= t_1(\tau_0)\cdot\tau_2^2 + t_7(\tau_0)\cdot\tau_2\cdot(\tau_1-\tau_0\tau_3) + t_9(\tau_0)\cdot\tau_0\cdot\tau_2 - \tau_2, \\ \hat{t}_2(\frac{1}{\tau_2}) &= t_2(\tau_0)\cdot\tau_2^3 + t_6(\tau_0)\cdot\tau_2^2\cdot(\tau_1-\tau_0\tau_3) + t_8(\tau_0)\cdot\tau_0\cdot\tau_2^2, \\ \hat{t}_3(\frac{1}{\tau_2}) &= t_3(\tau_0)\cdot\tau_2^4 + t_5(\tau_0)\cdot\tau_2^3\cdot(\tau_1-\tau_0\tau_3), \\ \hat{t}_4(\frac{1}{\tau_2}) &= t_4(\tau_0)\tau_2^5, \\ \hat{t}_5(\frac{1}{\tau_2}) &= -t_5(\tau_0)\cdot\tau_2^2\cdot(\tau_1-\tau_0\tau_3), \\ \hat{t}_6(\frac{1}{\tau_2}) &= -t_6(\tau_0)\cdot\tau_2\cdot(\tau_1-\tau_0\tau_3) - t_8(\tau_0)\cdot\tau_0\cdot\tau_2. \end{aligned}$$

4.15 All solutions of $\mathsf{R}_0,\ \check{\mathsf{R}}_0$

In this section we describe all the solutions of the non linear differential equations $\mathsf{R}_0,\ \check{\mathsf{R}}_0$ in terms of the solutions of the linear differential equation (2.25).

A solution of (2.25) can be written as $\int_{\delta_z}\eta$, where η is a holomorphic 3-form on $\mathsf{X}_{1,z}$ and δ_z is a continuous family of 3-cycles with complex coefficients. From this we only use the bilinear map $\langle\cdot,\cdot\rangle$ in the solution space of (2.25) induced by the intersection form in the third homology group of $\mathsf{X}_{1,z}$. Let $y_{i1},\ i=1,2$ be two linearly independent solutions of (2.25). The mirror map or Schwarz map is defined to be the multi valued function

$$\tau_0 = \frac{y_{11}}{y_{21}}. \tag{4.24}$$

Let

$$y_{ij} = \theta^{j-1} y_{i1},\ \ i=1,2,\ \ j=1,2,3,4$$

and t_i be the same expressions as in (2.27) replacing x_{ij} with y_{ij}.

Theorem 10 *If $\langle y_{11},y_{21}\rangle = 0$ then t_i's seen as functions of τ_0 form a solution of the vector field R_0. If $\langle y_{11},y_{21}\rangle = 1$ then t_i's seen as functions of τ_0 form a solution of the vector field $\check{\mathsf{R}}_0$.*

If we multiply periods y_{2j}'s with a constant a and y_{1j}'s with another constant b, then the resulting expressions for t_i's satisfy again the differential equation R_0. For $\check{\mathsf{R}}_0$ this is still true if we assume that $ab=1$, because we must keep the condition $\langle y_{11},y_{21}\rangle = 1$.

Proof. The proof for this theorem is purely computational. The first six equalities are just trivial identities. In order to check the last one we need to know which polynomial relations may exists between periods y_{ij}. We fix a symplectic basis x_{i1} of the linear differential (2.25), that is, $[\langle x_{i1},x_{j1}\rangle]$ is the 4×4 matrix Ψ defined in (0.1). We further assume that $x_{11}=y_{11}, x_{21}=y_{21}$ if $\langle y_{11},y_{21}\rangle=0$ and $x_{21}=y_{11}, x_{41}=y_{21}$ if $\langle y_{11},y_{21}\rangle = 1$. Now, for $\langle y_{11},y_{21}\rangle = 0$ we proceed as in the proof of Theorem 4. The other case is also similar and follows from the arguments in §4.12.

4.16 Around the elliptic point

We need the behavior of the quantities (2.27) at $z=\infty$. The monodromy of the linear differential (2.25) around $z=\infty$ is of order 5. We call this an elliptic point. We use the change of variable

$$\psi := z^{-\frac{1}{5}}.$$

Therefore $\theta = -\frac{1}{5}\psi\frac{\partial}{\partial\psi}$. This is the classical ψ-parameter mainly used in the Physics literature. The pair mainly used in Physics literature is $(\mathsf{X}_{1,z},5\omega_1)\cong(\mathsf{X}_{\psi,1},5\psi\omega_1)$. We conclude that the period x_{21} has a zero of order one at $\psi=0$. After some work we get the following expressions for t_i's:

$$
\begin{aligned}
t_0 &= \psi y_{21},\\
t_1 &= 5^4 y_{21}(5\psi^3 y_{22}+5\psi^4 y_{23}+(\psi^5-1)y_{24}),\\
t_2 &= 5^4 y_{21}^2(-5\psi^3 y_{21}+(-\psi^5+1)y_{23}),\\
t_3 &= 5^4 y_{21}^3(-5\psi^4 y_{21}+(\psi^5-1)y_{22}),\\
t_4 &= y_{21}^5,\\
t_5 &= 5^4 y_{21}^2(\psi^5-1)(y_{12}y_{21}-y_{11}y_{22}),\\
t_6 &= 5^4 y_{21}(\psi^5-1)(y_{11}y_{23}-y_{13}y_{21})
\end{aligned}
$$

[Supp Item 9]. Here $y_{i1} := \frac{x_{i1}}{\psi}$ and $y_{ij} = \frac{\partial^{j-1} y_{i1}}{\partial \psi^{j-1}}, \quad j = 1,2,3,4$. Note that y_{i1} are periods of $(\mathsf{X}_{1,\psi}, \omega_1)$ and so they are holomorphic at $\psi = 0$.

4.17 Halphen property

All solutions of R_0 and $\check{\mathsf{R}}_0$ can be constructed using Theorem 10. In this section we reformulate this in terms of the so-called Halphen property. This property for Darboux-Halphen differential equations was discovered by Halphen in [Hal81], see also [Mov12b]. For simplicity we work with R_0. Similar statements are also valid for $\check{\mathsf{R}}_0$.

Let $\mathbb{H}$ be the monodromy covering attached to mirror quintic Calabi-Yau varieties, see §4.6. For $w = (X, \delta) \in \mathbb{H}$ and $A \in \mathrm{Sp}(4, \mathbb{C})$, there is a unique basis α of $H^3_{\mathrm{dR}}(X)$, such that (X, α) is an element in the moduli space T and the period matrix

$$[\int_{A(\delta)_i} \alpha_j] = A[\int_{\delta_i} \alpha_j]$$

of the triple $(X, A\delta, \alpha)$ is of the τ-form. In this way we have well-defined meromorphic maps

$$t_A : \mathbb{H} \to \mathsf{T}, \quad \tau_A : \mathbb{H} \to \Pi$$

which are characterized by the uniqueness of the basis α and the equality

$$\tau_A(w) = [\int_{(A\delta)_i} \alpha_j].$$

Note that t_A and τ_A have poles on $w \in \mathbb{H}$ such that for $\mathsf{P} = A \cdot \tau(w)$ either $\mathsf{P}_{12} = 0$ or $\mathsf{P}_{11}\mathsf{P}_{22} - \mathsf{P}_{12}\mathsf{P}_{21} = 0$. For $A = I_{4\times 4}$ the identity matrix, we get $t := t_A$ and $\tau := \tau_A$ described in §4.6. We have also a map

$$\mathrm{Sp}(4, \mathbb{C}) \times \mathbb{H} \to \mathsf{G}, \ (A, w) \mapsto \mathsf{g}(A, w),$$

which satisfies

$$t(w) = t_A(w) \bullet \mathsf{g}(A, w),$$

$$A \cdot \tau(w) = \tau_A(w) \cdot \mathrm{g}(A, w).$$

For the identity matrix $I = I_{4\times4}$ we have $\mathrm{g}(I,w) = I$. If in $\mathbb{H}$ we consider the coordinate system τ_0, which is the $(1,1)$ entry of $\tau(w)$, we can calculate the maps $\tau_A(w), \mathrm{g}(A,w)$ explicitly. Since writing them in a mathematical paper, does not seem to be reasonable, we just write down the $(1,1)$ entry of $\tau_A(w)$:

$$\tilde{\tau}_0 := \tau_A(w)_{11} := \frac{a_{11}\tau_0(w) + a_{12} + a_{13}\tau_1(w) + a_{14}\tau_2(w)}{a_{21}\tau_0(w) + a_{22} + a_{23}\tau_1(w) + a_{24}\tau_2(w)},$$

where $A = [a_{ij}]_{4\times4} \in \mathrm{Sp}(4,\mathbb{C})$. The map t_A is tangent to R_0 but it does not make sense to say that it is a solution of R_0 because so far we have not fixed a coordinate system in $\mathbb{H}$. A reformulation of Theorem 10 is:

Theorem 11 *The meromorphic map $t_A : \mathbb{H} \to T$ with the coordinate system $\tilde{\tau}_0$ is a solution of* R_0. *In other words, for any $A \in \mathrm{Sp}(4,\mathbb{C})$,*

$$t_A(\tilde{\tau}_0) = t(A(\tilde{\tau}_0)) \bullet \mathrm{g}(A, \tilde{\tau}_0)$$

is a solution of R_0.

Proof. For the proof use A^{-1} instead of A. Note that we have redefined A to be A^{-1}.

We have $\dim\mathrm{Sp}(4,\mathbb{C}) = 10$ and the solution space of R_0 in T is of dimension 6. Therefore, the solution $t : \mathbb{H} \to T$ is fixed by a 4-dimensional subgroup of $\mathrm{Sp}(4,\mathbb{C})$.

4.18 Differential Calabi-Yau modular forms around the conifold

The indicial equation of the linear differential equation (2.25) at $z = 1$ is $\lambda(\lambda-1)^2(\lambda-2)$. We can therefore use the Frobenius method and calculate four unique and linearly independent solutions of the form

$$\begin{pmatrix}\phi_0\\ \phi_1\\ \phi_2\\ \phi_3\end{pmatrix} = \begin{pmatrix} O(z_1^5) + \frac{2971}{468750}z_1^5 + \frac{97}{18750}z_1^4 + \frac{2}{625}z_1^3 + 1 \\ O(z_1^5) + \frac{6089}{15625}z_1^5 + \frac{1133}{2500}z_1^4 + \frac{41}{75}z_1^3 + \frac{7}{10}z_1^2 + z_1 \\ O(z_1^3) + \phi_1 \cdot \ln(z_1) \\ O(z_1^5) + \frac{286471}{225000}z_1^5 + \frac{2309}{1800}z_1^4 + \frac{37}{30}z_1^3 + z_1^2 \end{pmatrix}, z_1 := (1-z). \quad (4.25)$$

We can calculate the coefficient of $(1-z)^i$, $i \geq 2$ by substituting ϕ_i with unknown coefficients in the differential equation [Supp Item 10]. We are interested in the analytic continuations of the periods x_{i1}, $i = 1,2,3,4$ in §2.7 along the interval $(0,1)$ and around the conifold singularity $z = 1$. Let us write the base change matrix

$$\begin{pmatrix}x_{11}\\ x_{21}\\ x_{31}\\ x_{41}\end{pmatrix} = C\begin{pmatrix}\phi_0\\ \phi_1\\ \phi_2\\ \phi_3\end{pmatrix}, \quad C = \begin{pmatrix} * & * & 0 & * \\ * & * & * & * \\ * & * & 0 & * \\ 0 & * & 0 & 0\end{pmatrix}, \quad (4.26)$$

where ϕ_i's are given in (4.25). Some of the entries of the above matrix as a closed formula or a numeric data can be found in [HKQ09]. The reason for zeros entries is as follows. By the monodromy of (2.25) written in the basis x_{i1}, $i = 1,2,3,4$ we know that x_{41} is one valued around $z = 1$ and the anticlockwise monodromy of x_{21} around $z = 1$ is $x_{21} + x_{41}$. The indicial equation of the linear differential equation (2.25) at $z = 1$ is $\lambda(\lambda-1)^2(\lambda-2)$. By Frobenius method, we conclude that x_{41} is, up to multiplication by a constant, is the unique holomorphic solution of order one at $z = 1$ and $x_{21} = x_{41}\frac{\ln(z-1)}{2\pi i} + hol.$ is a logarithmic solution.

The period expressions (2.27) for the special choice of the mirror map $\frac{x_{11}}{x_{21}}$ has some nice properties around the conifold singularity $z = 1$.

Proposition 13 *All the period expressions t_i's in* (2.27) *have limits as z goes to* 1 *in a sector with vertex* 1.

Proof. Using the Frobenius basis as in (4.25), we know that all t_i's are of the form $f(z)\ln(z-1) + g(z)$, where f and g are holomorphic functions around $z = 1$. It is enough to prove that f has no constant term around $z = 1$, that is, the Taylor series of f at $z = 1$ starts with $(z-1)$. Below the equalities are up to multiplication with a constant and up to adding a holomorphic function around $z = 1$. The proposition follows from

$$\begin{aligned}
x_{21} &= ((z-1) - \frac{7}{10}(z-1)^2 + \cdots)\ln(z-1), \\
x_{22} &= z(1 - \frac{7}{5}(z-1) + \cdots)\ln(z-1) = (1 - \frac{2}{5}(z-1) + \cdots)\ln(z-1), \\
x_{23} &= (-\frac{2}{5} + \cdots)\ln(z-1) + \frac{1}{z-1}, \\
x_{24} &= *\ln(z-1) + \frac{*}{(z-1)} + \frac{-1}{(z-1)^2}
\end{aligned}$$

and the fact that x_{11} is holomorphic around $z = 1$. In the expression for x_{42}, $*$ means some holomorphic function around $z = 1$ and for others $\cdots$ means higher order terms. The statement for t_0, t_3, t_4, t_5 directly follows from the format of x_{21} and the their expressions.

We may write t_i's as periods in terms of the derivation $(z-1)\frac{\partial}{\partial(z-1)}$. For t_5, t_2 (resp. t_1) division by $(z-1)$ (resp. $(z-1)^2$) appears and so we do not see the proof of the above proposition directly [Supp Item 11].

4.19 Logarithmic mirror map around the conifold

In this section we consider the mirror map whose denominator is x_{41}. Recall that x_{41} vanishes at $z = 1$ [Supp Item 12]. Consider the mirror map

$$\theta_0 := \frac{\tilde{x}_{21}}{x_{41}} = \frac{x_{21} + *x_{11} + *x_{31} + *x_{41}}{x_{41}}, \quad \hat{q} := e^{2\pi i\theta_0}, \tag{4.27}$$

where $*$'s are parameters such that the denominator of θ_0 is $x_{41}\ln(z-1)+O((z-1)^2)$ (one parameter is still free). Let $\hat{t}_i$'s be the quantities attached to this mirror map as in (2.27). All $\hat{t}_i$'s are holomorphic functions around $z=1$ and in fact we have

$$
\begin{aligned}
\hat{t}_0 &= \hat{a}\cdot(z_1+\cdots),\\
\hat{t}_1 &= \hat{a}^2\cdot(0z_1^2+\cdots),\\
\hat{t}_2 &= \hat{a}^3\cdot(0z_1^3+\cdots),\\
\hat{t}_3 &= \hat{a}^4\cdot(0z_1^4+\cdots),\\
\hat{t}_4 &= \hat{a}^5\cdot(z_1^5+\cdots),\\
\hat{t}_5 &= \frac{\hat{a}^4}{2\pi i}\cdot 5^5\cdot(z_1^4+\cdots)=\hat{a}^2\cdot(\frac{5^5\hat{a}^2}{2\pi i}z_1^4+\cdots),\\
\hat{t}_6 &= \frac{\hat{a}^3}{2\pi i}\cdot 5^5\cdot(*z_1^2+\cdots)=\hat{a}\cdot(*z_1^2+\cdots).
\end{aligned}
$$

All these follows from $x_{41}=\hat{a}\left(z_1+\frac{7}{10}z_1^2+\frac{41}{75}z_1^3+\frac{1133}{2500}z_1^4+\frac{6089}{15625}z_1^5+O(z_1^6)\right)$. Using Proposition 6, we know that

$$
\begin{aligned}
\hat{a}^{-1}\hat{t}_0 &= 1q+t_{0,2}q^2+\cdots+t_{0,n+1}q^{n+1}+\cdots,\\
\hat{a}^{-2}\hat{t}_1 &= 0q^2+t_{1,3}q^3+\cdots+t_{1,n+2}q^{n+2}+\cdots,\\
\hat{a}^{-3}\hat{t}_2 &= 0q^3+t_{2,4}q^4+\cdots+t_{2,n+3}q^{n+3}+\cdots,\\
\hat{a}^{-4}\hat{t}_3 &= 0q^4+t_{3,5}q^5+\cdots+t_{3,n+4}q^{n+4}+\cdots,\\
\hat{a}^{-5}\hat{t}_4 &= 1q^5+t_{4,6}q^6+\cdots+t_{4,n+5}q^{n+5}+\cdots,\\
\hat{a}^{-2}\hat{t}_5 &= \frac{5^5}{a}q^4+t_{5,5}q^5+\cdots+t_{5,n+4}q^{n+4}+\cdots,\\
\hat{a}^{-1}\hat{t}_6 &= 0q+t_{6,2}q^2+\cdots+t_{6,n+1}q^{n+1}+\cdots
\end{aligned}
$$

form a solution of $\check{\mathsf{R}}_0$ with

$$\dot{x}=aq\frac{\partial}{\partial q},\ a=\frac{2\pi i}{\hat{a}^2}.$$

We substitute them in $\check{\mathsf{R}}_0$. In the line 1, 2, 3, 4, 5, 6, 7 we compare the coefficients of

$$q^{n+5},\ q^{n+6},\ q^{n+7},\ q^{n+8},\ q^{n+9},\ q^{n+7},\ q^{n+5},$$

respectively. The coefficients

$$p_n:=(t_{0,n+1},\ t_{1,n+2},\ t_{2,n+3},\ t_{3,n+4},\ t_{4,n+5},\ t_{5,n+4},\ t_{6,n+1})$$

are those with the highest index in their respective group. In level $n=1$ we get $t_{5,4}=\frac{5^5}{a}$ and

$$(t_{0,2},t_{1,3},t_{2,4},t_{3,5},t_{4,6},t_{5,5},t_{6,2})$$

$$= (\frac{1}{9375}at_{5,5} + \frac{1}{2}t_{6,2}, -\frac{625}{2}t_{6,2}, -\frac{9375}{2}t_{6,2}, -\frac{15625}{2}t_{6,2}, \frac{1}{1875}at_{5,5} - \frac{5}{2}t_{6,2}, t_{5,5}, t_{6,2}).$$

Therefore, in this level we cannot compute $t_{5,5}$ and $t_{6,2}$. In fact, in the next level $n = 2$ we compute $t_{6,2}^2 = -\frac{5^5}{a}$. The recursion given by $\check{\mathsf{R}}_0$ in this case is not so easy to describe.

4.20 Holomorphic mirror map

Let us now consider the case in which both the denominator and nominator of the mirror map (4.24) are holomorphic functions around the singularity $z = 1$. This cannot happen around the maximal unipotent singularity $z = 0$ because there we have only one holomorphic solution (up to multiplication by a constant). The case of a mirror map with logarithmic nominator is discussed in §4.19.

The most general mirror map with holomorphic nominator and denominator that we can take is

$$\tau_0 = \frac{a_1x_{11} + a_2x_{31} + a_3x_{41}}{b_1x_{11} + b_2x_{31} + b_3x_{41}}$$

where $a_i, b_i \in \mathbb{C}$ are parameters and x_{ij}'s are as in §2.7. Note that x_{21} has logarithmic expression at $z = 1$ and the intersection of the nominator and denominator of τ_0 is $a_1b_2 - b_1a_2$. We further assume that the nominator, respectively denominator, is of the form $O((z-1)^a)$, respectively $O((z-1)^{a-1})$ for some $a \in \mathbb{N}$. In this way the mirror map is holomorphic at $z = 1$ and it vanishes at $z = 1$. We use it as a coordinate system around $z = 1$. Looking at the Frobenius basis around the conifold singularity $z = 1$, we realize that there exist only two possibilities for a, namely $a = 1, 2$.

Case $a = 2$: The mirror map depends on 2 parameters. For this mirror map I did not find any physics literature and so I will not discuss it here. However, note that the coordinates t_i of the corresponding solution of R_0 or $\check{\mathsf{R}}_0$ given by Theorem 4 has non-zero vanishing order at $z = 1$. The particular case $b_1 = b_2 = 0$ has many similarities with the logarithmic mirror map around $z = 1$, see §4.19. Note that $a_1b_2 - b_1a_2 = 0$ and so by Theorem 4 we get a solution of R_0.

Case $a = 1$: The mirror map depends on 4 parameters (2 parameters in the nominator and 3 parameters in the denominator and subtracting 1 because of the quotient). Using Theorem 4 we get a solution of R_0 for $a_1b_2 - a_2b_1 = 0$, respectively $\check{\mathsf{R}}_0$ for $a_1b_2 - a_2b_1 \neq 0$.

For $a_1b_2 - a_2b_1 = 0$ if we set $t_i := \sum_n t_{i,n}q^i$ and $\dot{x} = \frac{\partial x}{\partial \tau_0}$ and substitute in R_0 we get the recursion explained in 5.4. The particular case of this situation $a_1 = a_2 = 0, a_3 = 1$

$$\tau_0 := \frac{x_{41}}{b_1x_{11} + b_2x_{31} + b_3x_{41}}$$

is used in [HKQ09]. The intersection of the nominator and denominator is zero and so it result in a solution of R_0.

Chapter 5
Formal power series solutions

In this chapter we give some formal power series solutions of R_0 and $\check{\mathsf{R}}_0$. They converge to expressions which can be written in terms of the solutions of the Picard-Fuchs equation of mirror quintic Calabi-Yau threefolds, see §4.15. We would like to emphasize that it is possible to do Fourier or q-expansions using R_0 or $\check{\mathsf{R}}_0$ and without using Picard-Fuchs equations or the underlying geometries. For many examples of computing q-expansions in the same style as of this chapter see the author's webpage [Supp Item 13].

5.1 Singularities of modular differential equations

We are interested in solutions of the vector fields R_0 and $\check{\mathsf{R}}_0$, which are transversal to the singular locus of $t_5 \cdot R_0$ and $t_5\check{\mathsf{R}}_0$, respectively. The singular locus of a vector field is the set of points p such that all components of the vector field evaluated at p vanish.

The vector field R_0 is tangent to the hypersurfaces $t_4 = 0$ and $t_4 - t_0^5 = 0$. This follows from the equalities

$$
\begin{aligned}
d(t_4 - t_0^5)(\mathsf{R}_0) &= (t_4 - t_0^5)(\frac{18750t_0^4 + 5t_3}{t_5}), \\
dt_4(\mathsf{R}_0) &= t_4(\frac{15625t_0^4 + 5t_3}{t_5}).
\end{aligned}
\tag{5.1}
$$

The vector field $t_5 \cdot \mathsf{R}_0$ has the following singularities:

$$
Sin_0 : t_1 + 625t_0^2 = t_2 + 4375t_0^3 = t_3 + 3750t_0^4 = t_4 = t_0t_6 - 3t_5 = 0, \tag{5.2}
$$
$$
Sin_1 : t_5 = t_4 - t_0^5 = t_3 + 3125t_0^4 = 0, \tag{5.3}
$$
$$
Sin_2 : t_0 = t_2 = t_3 = t_4 = 0 \tag{5.4}
$$

[Supp Item 14]. The singular locus of the vector field $t_5\check{\mathsf{R}}_0$ is $Sin_1 \cup Sin_2$.

5.2 q-expansion around maximal unipotent cusp

Let us assume that the formal power series

$$t_i = \sum_{n=0}^{\infty} t_{i,n} q^n, \quad i = 0, 1, \ldots, 6$$

form a solution of R_0 with $\dot{x} = a \cdot q \cdot \frac{\partial x}{\partial q}$, where a and all $t_{i,j}$'s are unknown coefficients. Further, assume that the first coefficient of t_5 is not zero, that is, $t_{5,0} \neq 0$. Let

$$p_n := (t_{0,n}, t_{1,n}, \ldots, t_{6,n}).$$

We also use p_n as a 7×1 matrix. Comparing the coefficients of q^0 we know that p_0 is a singularity of R_0. We assume that p_0 is in Sin_0 in (5.2):

$$p_0 = (t_{0,0}, -625t_{0,0}^2, -4375t_{0,0}^3, -3750t_{0,0}^4, 0, t_{5,0}, \frac{3t_{5,0}}{t_{0,0}}) \in Sin_0,$$

where $t_{0,0}$ and $t_{5,0}$ are free parameters. Let us define

$$\check{\mathsf{R}}_0 = t_5 \mathsf{R}_0 = \sum_{i=0}^{6} \check{\mathsf{R}}_{0,i} \frac{\partial}{\partial t_i} \tag{5.5}$$

and define

$$\mathrm{jacob}(\check{\mathsf{R}}_0) := [\frac{\partial(\check{\mathsf{R}}_{0,i})}{\partial t_j}]_{i,j=0,1,\ldots,6}.$$

This is the Jacobian matrix of $\check{\mathsf{R}}_0$. Comparing the coefficients of q^1 we get the following equality:

$$(\mathrm{jacob}(\check{\mathsf{R}}_0)_{t=p_0} - a \cdot t_{5,0} I_{7\times 7}) p_1 = 0, \tag{5.6}$$

where I_7 is the 7×7 identity matrix. From this we get

$$\det(\mathrm{jacob}(\check{\mathsf{R}}_0)_{t=p_0} - a \cdot t_{5,0} I_{7\times 7}) = 0.$$

The above determinant is equal to $-(a \cdot t_{5,0})^6 (3125t_{0,0}^4 + a \cdot t_{5,0})$ and so

$$a = \frac{-3125t_{0,0}^4}{t_{5,0}}. \tag{5.7}$$

From (5.6) we also get

$$p_1 = \frac{t_{5,1}}{t_{5,0}} \left(\frac{-8t_{0,0}}{125}, 30t_{0,0}^2, \frac{1070t_{0,0}^3}{3}, \frac{355t_{0,0}^4}{3}, \frac{-5t_{0,0}^5}{3}, t_{5,0}, \frac{187t_{5,0}}{375t_{0,0}} \right),$$

where $t_{5,1}$ is another free parameter. Comparing the coefficients of q^n, $n \geq 2$ we get a recursion of type

$$(\text{jacob}(\check{\mathsf{R}}_0)+3125t_{0,0}^4\cdot nI_{7\times 7})p_n = \text{ polynomials in the entries of } p_i,\ i<n\ .$$

Note that

$$\det(\text{jacob}(\check{\mathsf{R}}_0)+3125t_{0,0}^4\cdot nI_{7\times 7}) = (5^5t_{0,0}^4)^7\cdot n^6\cdot(n-1)$$

and that we need $t_{0,0}, t_{5,0}$ and $t_{5,1}$ in order to calculate all $t_{i,n}$'s [Supp Item 15].

In §2.8 we have worked with the initial data

$$t_{0,0}=\frac{1}{5},\ t_{5,0}=-1,\ \ t_{5,1}=1875.$$

Let t_i be the solution obtained by setting $t_{0,0}=t_{5,0}=t_{5,1}=1$. Using Proposition 6, we conclude that the solution depending on three parameters $t_{0,0}, t_{5,0}$ and $t_{5,1}$ is of the form

$$(t_{0,0})^{i+1}\cdot t_i(\frac{t_{5,1}}{t_{5,0}}q),\quad i=0,1,2,\ldots,4,$$

$$t_{5,0}\cdot t_5(\frac{t_{5,1}}{t_{5,0}}q),\quad \frac{t_{5,0}}{t_{0,0}}\cdot t_6(\frac{t_{5,1}}{t_{5,0}}q).$$

Note that in the definition of the derivation we have to use a in (5.7).

In total, the formal power series, we described in this section depend on free parameters $t_{0,0}, t_{5,0}$ and $t_{5,1}$. We denote them by

$$t_{i,z=0},\ \ i=0,1,\ldots,6$$

in order to distinguish them with t_i's which are regular function on T.

5.3 Another q-expansion

Now, let us consider the case $t_{5,0}=0,\ t_{5,1}\neq 0$ with $\dot{x}=a\frac{\partial}{\partial q}$. Comparing the coefficients of q^0 we know that p_0 is again a singularity of R_0. We assume that

$$p_0=(t_{0,0},t_{1,0},t_{2,0},-3125t_{0,0}^4,t_{0,0}^5,0,t_{6,0})\in Sin_1,$$

where p_0 depends on 4 free parameters $t_{i,0},\ i=0,1,2,6$. Comparing the coefficients of q^1 we get

$$(\text{jacob}(\check{\mathsf{R}}_0)_{t=p_0}-a\cdot t_{5,1}I_{7\times 7})p_1=0,$$

where $\check{\mathsf{R}}_0$ is given in (5.5). We have

$$\det(\text{jacob}(\check{\mathsf{R}}_0)-a\cdot t_{5,1}I_{7\times 7}) = -(a\cdot t_{5,1}-3125t_{0,0}^4)^3(a\cdot t_{5,1})^4$$

and so

$$a=\frac{3125t_{0,0}^4}{t_{5,1}}.$$

The matrix $\mathrm{jacob}(\check{\mathsf{R}}_0)_{t=p_0} - 3125t_{0,0}^4 I_{7\times 7}$ is of rank 5 and so p_1 is in the $\mathbb{C}$-vector space generated by two vectors:

$$p_1^1 = (0,0,0,0,0,3125t_{0,0}^4, -3125t_{0,0}^3 - 2t_{2,0}),$$
$$p_1^2 = (t_{0,0}, t_{1,0}, 2t_{2,0}, -9375t_{0,0}^4, 5t_{0,0}^5, 0, 3t_{6,0}),$$

$$p_1 = \frac{t_{5,1}}{3125t_{0,0}^4} p_1^1 + \frac{t_{0,1}}{t_{0,0}} p_1^2.$$

Comparing the coefficients of q^n, $n \geq 2$ we get a recursion of type

$$\mathrm{jacob}(\check{\mathsf{R}}_0)p_n - a \cdot t_{5,1} \cdot n \cdot p_n - a \cdot t_{5,n} \cdot p_1 = \text{ polynomials in the entries of } p_i,\ i < n.$$

This is simplified into

$$(A - n \cdot 5^5 \cdot t_{0,0}^4 \cdot I_{7\times 7})p_n = \text{ polynomial in terms of the entries of } p_i,\ i < n,$$

where A in terms of its columns is written in the form

$$A := \mathrm{jacob}(\check{\mathsf{R}}_0) - [0,0,0,0,ap_1,0].$$

Note that we have

$$\det(A - n \cdot 5^5 \cdot t_{0,0}^4 \cdot I_{7\times 7}) = -(5^5 t_{0,0})^7 n^5 (n-1)^2.$$

The formal power series, which we described in this section, depends on 6 free parameters $t_{i,0}$, $i = 0,1,2,6, t_{0,1}, t_{5,1}$ [Supp Item 16].

Proposition 14 *The q-expansions attached to each t_i as above satisfy $t_4 - t_0^5 = 0$.*

Proof. We first observed this statement experimentally [Supp Item 17]. A proof can be given as follows. The vector field R_0 is tangent to the hypersurface $t_4 - t_0^5 = 0$. We make the change of variable $\tilde{t}_4 = t_4 - t_0^5$ and write $\tilde{\mathsf{R}}_0 := \mathsf{R}_0 |_{\tilde{t}_4 = 0}$ in the coordinates $t_0, t_1, t_2, t_3, t_5, t_6$. The whole discussion above works in this six-dimensional space.

5.4 q-expansion around conifold

Let us take a mirror map $q := \frac{y_{11}}{y_{21}}$ with $y_{11} = O((z-1)^1), y_{21} = O((z-1)^0)$ and around the conifold point $z = 1$, and write the corresponding q-expansions for t_i's. We have $t_4 - t_0^5 = (z-1)y_{21} = O((z-1)^1)$. The first coefficients vector $(t_{i,0},\ i = 0,1,\ldots,6)$ is not a generic point of Sin_1 as it was assumed in §5.3. This leads us to the following discussion which gives the true q-expansion around conifold.

In order to get q-expansions which converge and correspond to a mirror map around the conifold we choose

$$p_0 = (t_{0,0}, t_{1,0}, 3125t_{0,0}^3, -3125t_{0,0}^4, t_{0,0}^5, 0, 0) \in Sin_1,$$

which is a two-dimensional sublocus $Sin_{1,c}$ of Sin_1. The particular expressions for $t_{2,0}$ and $t_{6,0}$ follow from the period expressions of t_i in (2.27). The difference between this and a generic point $p_0 \in Sin_1$ appears when we calculate the coefficients of q^1. The matrix $\mathrm{jacob}(\check{\mathsf{R}}_0)_{t=p_0} - 3125t_{0,0}^4 I_{7\times 7}$ turns out to be of rank 4 (and not 5). The vector p_1 is in the $\mathbb{C}$-vector space generated by three vectors:

$$\begin{aligned}
p_1^1 &:= \Big(625t_{0,0}^2 + t_{1,0}, 0, -3906250t_{0,0}^4 - 9375t_{0,0}^2 t_{1,0} - t_{1,0}^2,\\
&\quad -5859375t_{0,0}^5 - 12500t_{0,0}^3 t_{1,0}, 3125t_{0,0}^6, 0, 0\Big),\\
p_1^2 &:= \Big(0, (625t_{0,0}^2 + t_{1,0}), (3125t_{0,0}^3 + t_{0,0}t_{1,0}), 3125t_{0,0}^4, 5t_{0,0}^5, 0, 0\Big),\\
p_1^3 &:= (0,0,0,0,0,t_{0,0},1)
\end{aligned}$$

and

$$p_1 = e_1 p_1^1 + e_2 p_1^2 + e_3 p_1^3,$$

where e_1, e_2, e_3 are additional free parameters. The structure of the recursion is as in §5.4. Note that in this case

$$a = \frac{3125t_{0,0}^4}{t_{5,1}} = \frac{3125t_{0,0}^3}{e_3}.$$

If we set $a = 1$ then in total we have four free parameters which is the same number of freedom for a mirror map around the conifold. We denote them by

$$t_{i,z=1}, \quad i = 0, 1, \ldots, 6$$

in order to distinguish them with t_i's which are regular function on T. For the computer codes used in the present section see the author's webpage [Supp Item 18].

5.5 New coordinates

We change the coordinates

$$\tilde{t}_4 = t_4 - t_0^5,\ \tilde{t}_3 = t_3 + 3125t_0^4,\ \tilde{t}_2 = t_2 + 3125t_0^3 \tag{5.8}$$

and for simplicity remove the tilde sign. The differential equation R_0 after the change of coordinates (5.8) turns out to be

$$\dot{t}_0 = \frac{1}{t_5}(t_3t_0 - 625t_4),$$
$$\dot{t}_1 = \frac{1}{t_5}(t_3t_1 + 390625t_4t_0),$$
$$\dot{t}_2 = \frac{1}{t_5}(3125t_3t_0^3 + 2t_3t_2 + 625t_4t_1),$$
$$\dot{t}_3 = \frac{1}{t_5}(3125t_3t_0^4 + 3t_3^2 + 625 \cdot t_4t_2),$$
$$\dot{t}_4 = \frac{1}{t_5}(3125t_4t_0^4 + 5t_4t_3),$$
$$\dot{t}_5 = \frac{1}{t_5}(3125t_5t_0^4 + 625t_6t_4 + 2t_5t_3),$$
$$\dot{t}_6 = \frac{1}{t_5}(3125t_5t_0^3 + 3t_6t_3 - 2t_5t_2)$$

[Supp Item 19]. The vector field $t_5\mathsf{R}_0$ has the singular locus Sin_1 given by $t_3 = t_4 = t_5 = 0$. Inside Sin_1 we are interested to do q-expansions around

$$Sin_{1,c}: \ t_2 = t_3 = t_4 = t_5 = t_6 = 0.$$

We write $t_i = \sum_{j=0} t_{0,j}q^n$ with $t_{2,0} = t_{3,0} = t_{4,0} = t_{5,0} = t_{6,0} = 0$ and plug it inside the above differential equation with $\dot{t} = a\frac{\partial t}{\partial q}$ and we get a recursion of all other coefficients. For this recursion, we consider the first initial values as parameters.

5.6 Holomorphic foliations

We can look for the holomorphic foliation $\mathscr{F}_0$ induced by R_0 in $\mathbb{C}^7$ and study its dynamics. The similar study for the Ramanujan vector field shows a rich arithmetic and dynamic structure, for more details see [Mov08b]. The equalities in (5.1) show that the hypersurfaces $t_4 = 0$ and $t_4 - t_0^5 = 0$ are $\mathscr{F}_0$-invariants. In these hypersurfaces we have two singular loci Sin_0 and Sin_1, respectively. The discussion in §5.2 implies that for any point $p \in Sin_0 - \{t_5 = t_0 = 0\}$ we have a leaf of $\mathscr{F}_0$ transverse to Sin_0 at p. In other words in a neighborhood of $Sin_0 - \{t_5 = t_0 = 0\}$ we have a dimension 3 analytic subvariety which is $\mathscr{F}_0$-invariant. A similar statement is valid for Sin_1. The discussion in §5.3 implies that for any point $p \in Sin_1 - \{t_0 = 0\}$ we have a one parameter family of leaves of $\mathscr{F}_0$ transverse to Sin_1 at p and inside the hypersurface $t_4 - t_0^5 = 0$. Since $\dim(Sin_0) = 4$, we conclude that inside $t_4 - t_0^5 = 0$ and near Sin_1 all the leaves are transversal to Sin_1. We can also describe a similar geometry for the discussion in §5.4. In all these cases the foliation $\mathscr{F}_0$ has radial type singularities along Sin_0 and Sin_1, respectively. The singularity Sin_2 is not used in our discussion of differential Calabi-Yau modular forms.

Chapter 6
Topological string partition functions

We describe a Lie Algebra on the moduli of mirror quintic Calabi-Yau threefolds enhanced with differential forms and its relation to Bershadsky-Cecotti-Ooguri-Vafa holomorphic anomaly equation. In particular, we describe the polynomial structure of topological partition functions $\mathsf{F}_g^{\mathrm{alg}}$, $g \geq 0$ in terms of seven quantities which are generalizations of quasi-modular forms. In this way, we recover a result of Yamaguchi-Yau in an algebraic geometric context.

6.1 Yamaguchi-Yau's elements

In order to reduce the confusion between our period expressions in Theorem 4, and those in Physics literature [BCOV93, BCOV94] and mainly [YY04], we have worked out Table 1 [Supp Item 20]. In this table whenever there is an equality sign $=$ it means that the quantities in the left and right hand side are equal. If there is not, then both quantities are not necessarily the same.

In [YY04], Yamaguchi and Yau prove that the topological string partition function $\mathsf{F}_g^{\mathrm{hol}}$ for the mirror quintic Calabi-Yau variety is of the form $W_1^{3g-3}P_g(v_1, v_2, v_3, X)$, where P_g is a polynomial of degree at most $3g-3$ with weights $\deg(v_i) = i$, $i = 1,2,3$, $\deg(X) = 1$ and W_1, v_1, v_2, v_3, X are given in the table below. The ambiguity in determining P_g comes from the terms

$$X^j \quad 0 \leq j \leq 3g-3.$$

Considering the weights (2.11) we have $\deg(X) = \deg(v_1) = \deg(v_2) = \deg(v_3) = 0$, $\deg(W_1^3) = 2$. By looking the pole and zero orders of X, v_1, v_2, v_3, W_1^3 at $(t_4 - t_0^5)$ and t_5 we get a proof of (2.40). Note that $\mathsf{F}_g^{\mathrm{hol}}$ has no poles at t_0 and this solves $g-1$ number of ambiguities. From now on, we use $\mathsf{F}_g^{\mathrm{alg}}$ to denote $\mathsf{F}_g^{\mathrm{hol}}$ written as a rational function in t_i's.

Name	Physics notation	My notation
Kähler parameter Mirror map	$t =$	$-2\pi i \tau_0$
The complex moduli	$\psi^{-5} =$	z
Logarithmic derivation	$\psi\partial_\psi =$	-5θ where $\theta = z\frac{\partial}{\partial z} = \frac{(x_{12}x_{21}-x_{11}x_{22}))}{x_{21}^2}\frac{\partial}{\partial \tau_0}$ $\frac{t_0 t_5}{5^5(t_4-t_0^5)}\frac{\partial}{\partial \tau_0}$
Mirror quintic variety Affine part	$W_\psi =$ $Q = x_1^5+x_2^5+x_3^5+x_4^5+x_5^5-5\psi x_1x_2x_3x_4x_5$	$W_{1,z}$ $F = -x_1^5-x_2^5-x_3^5-x_4^5-zx_5^5+5x_1x_2x_3x_4x_5$ $f = -x_1^5-x_2^5-x_3^5-x_4^5-z+5x_1x_2x_3x_4$
Holomorphic 3-forms	$\Omega = \frac{x_5 dx_1\wedge dx_2\wedge dx_3}{\frac{\partial Q}{\partial \psi}} =$	5η, where $\eta = \frac{dx_1\wedge d_2\wedge dx_3\wedge dx_4}{df}$
Yukawa coupling	$C_{\psi\psi\psi} = (\frac{2\pi i}{5})^3\frac{\psi^2}{1-\psi^5}$	$\frac{1}{1-z}$
Yamaguchi-Yau variable	$W_1^3 = \frac{\omega_0^2(\psi^{-5}-1)}{5^4}$	$\frac{t_4-t_0^5}{5^4 t_0^3}$
Yamaguchi-Yau variable	$X = \frac{1}{1-\psi^5}$	$\frac{t_4}{t_4-t_0^5}$
Yamaguchi-Yau variable	$v_1 = \frac{(\psi\partial_\psi)^{(2)}(t)}{(\psi\partial_\psi)(t)} + 2\frac{(\psi\partial_\psi)(\omega_0)}{\omega_0}$	$\frac{3t_5-t_0t_6}{t_5}$
Yamaguchi-Yau variable	$v_2 = \frac{(\psi\partial_\psi)^{(2)}\omega_0}{\omega_0} - v_1\frac{(\psi\partial_\psi)\omega_0}{\omega_0}$	$\frac{-3750t_0^6t_6+4375t_0^5t_5+t_0^2t_2t_5-t_0^2t_3t_6+625t_0t_4t_6-1250t_4t_5}{5^4t_5(t_4-t_0^5)}$
Yamaguchi-Yau variable	$v_3 = \frac{(\psi\partial_\psi)^{(3)}\omega_0}{\omega_0} - \frac{(\psi\partial_\psi)\omega_0}{\omega_0}(-v_2+5Xv_1-10X)$	v_3
TSPF genus 1	$\mathsf{F}_1^{\mathrm{hol}} = \frac{1}{2}\log\left(\left(\frac{\psi}{\omega_0}\right)^{\frac{62}{3}}(1-\psi^5)^{-\frac{1}{6}}\frac{\partial\psi}{\partial t}\right)$	$\frac{\partial}{\partial\tau_0}\ln\mathsf{F}_1^{\mathrm{alg}} = \frac{3750t_0^5t_6-353125t_0^4t_5-112t_3t_5-3750t_4t_6}{12t_5^2}$

$$
\begin{aligned}
v_3 = & \frac{1}{5^8(t_4-t_0^5)^2t_5}(14062500t_0^{11}t_6 - 14062500t_0^{10}t_5 + 625t_0^8t_1t_5 - 4375t_0^7t_2t_5 \\
& +7500t_0^7t_3t_6 - 3125t_0^6t_3t_5 - 16406250t_0^6t_4t_6 + 16406250t_0^5t_4t_5 \\
& -625t_0^3t_1t_4t_5 - t_0^3t_2t_3t_5 + t_0^3t_3^2t_6 + 4375t_0^2t_2t_4t_5 \\
& -16406250t_0^6t_4t_6 + 16406250t_0^5t_4t_5 - 625t_0^3t_1t_4t_5 - t_0^3t_2t_3t_5 \\
& +t_0^3t_3^2t_6 + 4375t_0^2t_2t_4t_5 - 4375t_0^2t_3t_4t_6 + 2343750t_0t_4^2t_6 - 2343750t_4^2t_5)
\end{aligned}
$$

6.2 Proof of Theorem 8

The proof is purely computational [Supp Item 21]. We have already constructed an isomorphism (2.8) and a universal family $\mathsf{X} \to \mathsf{T}$ of mirror quintics. By our construction we have global sections $\alpha_i \in H^3(\mathsf{X}/\mathsf{T})$, $i = 1,2,3,4$ and we have computed the Gauss-Manin connection of $\mathsf{X} \to \mathsf{T}$ written in the basis $\alpha := [\alpha_1, \alpha_2, \alpha_3, \alpha_4]^{\mathrm{tr}}$:

$$\nabla\alpha = \mathsf{A}\otimes\alpha.$$

Here A is a 4×4 matrix whose entries are explicit differential 1-forms in T. By Griffiths transversality the $(1,3),(1,4),(2,4)$ entries of A are zero. Having A explicitly, it is just a linear algebra to prove Theorem 8. Along the way, we have also computed explicitly the vector fields R_i and Y. The regular function Y turns out to be the Yukawa coupling (2.10) and

$$
\begin{aligned}
\mathsf{R}_0 = {} & \frac{3750t_0^5 + t_0t_3 - 625t_4}{t_5}\frac{\partial}{\partial t_0} - \frac{390625t_0^6 - 3125t_0^4t_1 - 390625t_0t_4 - t_1t_3}{t_5}\frac{\partial}{\partial t_1} \\
& - \frac{5859375t_0^7 + 625t_0^5t_1 - 6250t_0^4t_2 - 5859375t_0^2t_4 - 625t_1t_4 - 2t_2t_3}{t_5}\frac{\partial}{\partial t_2} \\
& - \frac{9765625t_0^8 + 625t_0^5t_2 - 9375t_0^4t_3 - 9765625t_0^3t_4 - 625t_2t_4 - 3t_3^2}{t_5}\frac{\partial}{\partial t_3} \\
& + \frac{15625t_0^4t_4 + 5t_3t_4}{t_5}\frac{\partial}{\partial t_4} - \frac{625t_0^5t_6 - 9375t_0^4t_5 - 2t_3t_5 - 625t_4t_6}{t_5}\frac{\partial}{\partial t_5} \\
& + \frac{9375t_0^4t_6 - 3125t_0^3t_5 - 2t_2t_5 + 3t_3t_6}{t_5}\frac{\partial}{\partial t_6},
\end{aligned}
$$

$$\mathsf{R}_1 = t_5\frac{\partial}{\partial t_5} + t_6\frac{\partial}{\partial t_6},$$

$$\mathsf{R}_2 = t_0\frac{\partial}{\partial t_0} + 2t_1\frac{\partial}{\partial t_1} + 3t_2\frac{\partial}{\partial t_2} + 4t_3\frac{\partial}{\partial t_3} + 5t_4\frac{\partial}{\partial t_4} + 3t_5\frac{\partial}{\partial t_5} + 2t_6\frac{\partial}{\partial t_6},$$

$$\mathsf{R}_3 = -\frac{5t_0^4t_6 - 5t_0^3t_5 - \frac{1}{625}t_2t_5 + \frac{1}{625}t_3t_6}{t_0^5 - t_4}\frac{\partial}{\partial t_1} + t_6\frac{\partial}{\partial t_2} + t_5\frac{\partial}{\partial t_3},$$

$$\mathsf{R}_4 = \frac{625t_0^5 - 625t_4}{t_5}\frac{\partial}{\partial t_6},$$

$$\mathsf{R}_5 = \frac{-3125t_0^4 - t_3}{t_5}\frac{\partial}{\partial t_1} + \frac{625(t_0^5 - t_4)}{t_5}\frac{\partial}{\partial t_2},$$

$$\mathsf{R}_6 = \frac{\partial}{\partial t_1}.$$

6.3 Genus 1 topological partition function

According to [BCOV93] the genus one topological partition function is given by

$$\mathsf{F}_1^{\text{hol}} = -\frac{1}{2}\log\left((\theta\frac{x_{11}}{x_{21}})^{-1}x_{21}^{-3-h^{11}+\frac{\chi}{12}}z^{a_0}(1-z^{-1})^{a_1}\right). \tag{6.1}$$

Here a_0 and a_1 are two ambiguities and

$$\chi = -200, \ \ h^{11} = 1.$$

Assume that the derivation of $\mathsf{F}_1^{\text{hol}}$ with respect to τ_0 is holomorphic at $z = \infty$. Writing in t_i's, this is equivalent to say that $\mathsf{R}_0\mathsf{F}_1^{\text{alg}}$ as a rational function in t_i's has no poles in $t_0 = 0$ [Supp Item 22]. In this way we get $a_0 = -13/3$ and

$$\mathsf{R}_0\mathsf{F}_1^{\text{alg}} = -\frac{1875t_0^5t_6 + 9375t_0^4t_5a_1 - 175000t_0^4t_5 - 56t_3t_5 - 1875t_4t_6}{6t_5^2}.$$

The ambiguity a_1 must be solved by studying $\mathsf{F}_1^{\mathrm{alg}}$ at $z=1$. In [BCOV93] we find that the asymptotic of $\mathsf{F}_1^{\mathrm{hol}}$ near $z=1$ is $\frac{1}{12}\ln(z-1)$ and so $a_1=-\frac{1}{6}$. Therefore

$$\mathsf{R}_0\mathsf{F}_1^{\mathrm{alg}}=-\frac{3750t_0^5t_6-353125t_0^4t_5-112t_3t_5-3750t_4t_6}{12t_5^2}.$$

After computing the q-expansion of $\mathsf{F}_1^{\mathrm{hol}}$ we get

$$\begin{aligned}q\frac{\partial}{\partial q}\mathsf{F}_1^{\mathrm{hol}}&=\frac{25}{12}-\sum_{n,r=1}\frac{n\cdot r\cdot d_rq^{nr}}{1-q^{nr}}-\frac{1}{12}\sum_{s=1}^{\infty}\frac{sn_sq^s}{(1-q^s)}\\&=\frac{25}{12}-\sum_{n=1}\Big(\sum_{r|n}d_r\Big)\frac{nq^n}{1-q^n}-\frac{1}{12}\sum_{s=1}^{\infty}\frac{sn_sq^s}{(1-q^s)}\end{aligned}$$

[Supp Item 23]. Here, n_d is the virtual number of rational curves in a generic quintic threefold

$$(n_d)_{d\in\mathbb{N}}=(2875,609250,317206375,242467530000,229305888887625,\cdots)$$

[Supp Item 24] and d_r is the virtual number of elliptic curves of degree r in a generic quintic threefold

$$(d_r)_{r\in\mathbb{N}}=(0,0,609250,3721431625,12129909700200,31147299733286500,\cdots)$$

[Supp Item 25]. The numbers n_s and d_r are also called the geneus zero and genus one instanton numbers, respectively. The ambiguity a_1 is fixed if, for instance, we prove that $d_1=0$ or $d_2=0$. Finally, note that

$$\mathsf{F}_1^{\mathrm{alg}}=\ln\big(t_4^{\frac{25}{12}}(t_4-t_0^5)^{\frac{-5}{12}}t_5^{\frac{1}{2}}\big) \tag{6.2}$$

and

$$t_4^{\frac{25}{12}}(t_4-t_0^5)^{-\frac{5}{12}}t_5^{\frac{1}{2}}=5^{-\frac{5}{12}}q^{\frac{25}{12}}\Big(\prod_{n=1}^{\infty}(1-q^n)^{\sum_{r|n}d_r}\Big)\Big(\prod_{s=1}^{\infty}(1-q^s)^{n_s}\Big)^{\frac{1}{12}}. \tag{6.3}$$

Note that R_0 translated into the holomorphic context and in q coordinate is $5q\frac{\partial}{\partial q}$.

6.4 Holomorphic anomaly equation

The Yukawa coupling is by definition the quantity (2.10). It satisfies the following equalities with respect to the vector fields R_i:

$$\begin{aligned} \mathsf{R}_1(\mathsf{Y}) &:= -3\mathsf{Y}, \\ \mathsf{R}_2(\mathsf{Y}) &:= \mathsf{Y}, \\ \mathsf{R}_i(\mathsf{Y}) &:= 0, \quad i = 3,4,5,6. \end{aligned} \tag{6.4}$$

These are used in the calculation of the Lie algebra structure of R_i's. Vice versa, we can derive the above equalities from the Lie algebra structure of R_i's. For instance, knowing that $[\mathsf{R}_1, \mathsf{R}_0] = -\mathsf{R}_0$ and Theorem 8 we can derive $\mathsf{R}_1\mathsf{Y} = -3\mathsf{Y}$.

The genus one topological partition $\mathsf{F}_1^{\text{alg}}$ by definition is (2.38). It satisfies the following equalities

$$\begin{aligned} \mathsf{R}_1\mathsf{F}_1^{\text{alg}} &= \frac{1}{2}, \\ \mathsf{R}_2\mathsf{F}_1^{\text{alg}} &= -\frac{1}{2}(c-1), \\ \mathsf{R}_i\mathsf{F}_1^{\text{alg}} &= 0, \ i = 3,4,5,6, \end{aligned} \tag{6.5}$$

where

$$c = -1 - h^{11} + \frac{\chi}{12}, \ h^{11} = 1, \ \chi := -200. \tag{6.6}$$

From the table (7) it follows that

$$\mathsf{R}_4\mathsf{R}_0\mathsf{F}_1^{\text{alg}} = -\frac{1}{2}\mathsf{Y}.$$

This can be considered as analog of the anomaly equation for $\mathsf{F}_1^{\text{alg}}$.

Let us define the formal power series

$$Z := \exp \sum_{g=1}^{\infty} \mathsf{F}_g^{\text{alg}} \lambda^{2g-2}. \tag{6.7}$$

The master anomaly equation in the language of the moduli space T and vector fields R_i in T is the following collection of equalities:

$$\begin{aligned} \mathsf{R}_1 Z &= \frac{1}{2} Z, \\ \mathsf{R}_2 Z &= (-\frac{1}{2}(c-1) + \lambda \partial_\lambda) Z, \\ \mathsf{R}_3 Z &= 0, \\ \mathsf{R}_4 Z &= \frac{1}{2}\lambda^2 \mathsf{R}_0^2 Z, \\ \mathsf{R}_5 Z &= -\frac{1}{2}\lambda^2((2-c) + 2\lambda\partial_\lambda)\mathsf{R}_0 Z, \\ \mathsf{R}_6 Z &= \frac{1}{4}\lambda^2(2-c+2\lambda\partial_\lambda)(-\frac{1}{2}c + \lambda\partial_\lambda) Z, \end{aligned} \tag{6.8}$$

where c is given in (6.6). The first three equalities in (6.8) follow form (6.5) and (2.39). The fourth equality is the reformulation of the holomorphic anomaly equation in (2.39). Note that $\mathsf{R}_4\mathsf{F}_1^{\mathrm{alg}} = 0$ and so the holomorphic anomaly equation has no extra term. The fifth and sixth equalities follow from the Lie algebra structure of R_i's and the fourth equality. Below, the equalities are written up to their action on Z, for instance, $\mathsf{R}_3 = 0$ means $\mathsf{R}_3 Z = 0$.

$$
\begin{aligned}
0 &= \mathsf{R}_3(\mathsf{R}_4 - \frac{1}{2}\lambda^2\mathsf{R}_0^2) \\
&= [R_3, \mathsf{R}_4] + \mathsf{R}_4\mathsf{R}_3 - \frac{1}{2}\lambda^2([\mathsf{R}_3, \mathsf{R}_0] + \mathsf{R}_0\mathsf{R}_3)\mathsf{R}_0 \\
&= -R_5 - \frac{1}{2}\lambda^2(\mathsf{R}_2 - \mathsf{R}_1 + \mathsf{R}_0\mathsf{R}_3)\mathsf{R}_0 \\
&= -R_5 - \frac{1}{2}\lambda^2([\mathsf{R}_2, \mathsf{R}_0] + \mathsf{R}_0\mathsf{R}_2 - [\mathsf{R}_1, \mathsf{R}_0] - \mathsf{R}_0\mathsf{R}_1 + \mathsf{R}_0([\mathsf{R}_3, \mathsf{R}_0] + \mathsf{R}_0\mathsf{R}_3)) \\
&= -R_5 - \frac{1}{2}\lambda^2(\mathsf{R}_0 + \mathsf{R}_0\mathsf{R}_2 + \mathsf{R}_0 - \mathsf{R}_0\mathsf{R}_1 + \mathsf{R}_0(\mathsf{R}_2 - \mathsf{R}_1)) \\
&= -R_5 - \frac{1}{2}\lambda^2((2-c)\mathsf{R}_0 + 2\lambda\partial_\lambda \mathsf{R}_0), \\
0 &= \mathsf{R}_3(R_5 + \frac{1}{2}\lambda^2(2 - c + 2\lambda\partial_\lambda)\mathsf{R}_0) \\
&= [\mathsf{R}_3, \mathsf{R}_5] + \mathsf{R}_5\mathsf{R}_3 + \frac{1}{2}\lambda^2(2 - c + 2\lambda\partial_\lambda)([\mathsf{R}_3, \mathsf{R}_0] + \mathsf{R}_0\mathsf{R}_3) \\
&= -2\mathsf{R}_6 + \frac{1}{2}\lambda^2(2 - c + 2\lambda\partial_\lambda)(\mathsf{R}_2 - \mathsf{R}_1) \\
&= -2\mathsf{R}_6 + \frac{1}{2}\lambda^2(2 - c + 2\lambda\partial_\lambda)(-\frac{1}{2}c + \lambda\partial_\lambda).
\end{aligned}
$$

6.5 Proof of Proposition 1

If we substitute each t_i by its q-expansion around the conifold singularity, as it is described in §5.4, then we must have a pole order $2g-2$. Since all $t_2 + 3125t_0^3, t_3 + 3125t_0^4, t_4 - t_0^5, t_5, t_6$ vanish at $q = 0$, and $t_5^{3g-3}(t_4 - t_0^5)^{2g-2}$ has a zero order $5g-5$ at $q = 0$, Q_g must have zero order $3g-3$ at $q = 0$ and we get the desired restriction on the monomials of Q_g.

6.6 The ambiguity of $\mathsf{F}_g^{\mathrm{alg}}$

If $\mathsf{F}_g^{\mathrm{alg}}$ is a solution of (2.39) then $\mathsf{F}_g^{\mathrm{alg}}$ plus any function in t_0, t_4 is also a solution. Using the asymptotic behavior of $\mathsf{F}_g^{\mathrm{alg}}$'s in [BCOV93], we know that the ambiguities of $\mathsf{F}_g^{\mathrm{alg}}$ arise from the coefficients of

$$\frac{P_g(t_0,t_4)}{(t_4-t_0^5)^{2g-2}}, \quad \deg(P_g) = 36(g-1). \tag{6.9}$$

Knowing that we are using the weights (2.11), we observe that it depends on $[\frac{12(g-1)}{5}]+1$ coefficients. The monomials in (6.9) are divided into two groups, those meromorphic in $t_4 - t_0^5$ and the rest which is

$$t_0^a(t_4-t_0^5)^b, \quad a+5b = 2g-2, \; a,b \in \mathbb{N}_0.$$

The coefficients of the first group can be fixed by the so-called gap condition and the asymptotic behaviour of $\mathsf{F}_g^{\mathrm{alg}}$ at the conifold, see §11.11. One of the coefficients in the second group can be solved using the asymptotic behaviour of $\mathsf{F}_g^{\mathrm{alg}}$ at the maximal unipotent monodromy. In total, we have $[\frac{2g-2}{5}]$ ambiguities which we do not know if it is possible to solve them using any data attached to mirror quintic Calabi-Yau threefold. Using the generating function role that $\mathsf{F}_g^{\mathrm{alg}}$'s play for counting curves in a generic quintic (A-model Calabi-Yau threefold), we may solve all the ambiguities, however, such computations are usually hard to perform. There is one exception to this due to Castelnuevo's theorem. Let C be a smooth curve of genus g and degree d in the projective space $\mathbb{P}^n$. Then $g \leq (n-1)\frac{m(m-1)}{2} + \varepsilon$, where m and ε are the quotient and remainder of the division of $d-1$ by $n-1$, for more details see [GH94] page 533. In particular, if we fix the genus g then we get a lower bound for the degree of the curve C. For $n = 4$ this can be used in order to fix the ambiguity of $\mathsf{F}_g^{\mathrm{alg}}$ up to genus 51, see for instance [HKQ09].

Remark 5 We may try to realize T as an affine scheme over $\mathbb{Z}[\frac{1}{N}]$ for some integer N (most probably $N = 5$). In this way $\mathsf{F}_g^{\mathrm{alg}}$'s have integral coefficients and so we may use original definition of $\mathsf{F}_g^{\mathrm{alg}}$'s by path integrals and approximate all the ambiguities.

Remark 6 Let us consider the non commutative ring $\mathbb{Q}[\mathsf{R}_i,\ i=0,1,\ldots,6,\lambda\partial_\lambda]$ and its left ideal I containing all elements which annihilate Z. The equalities (6.8) give us another left ideal $J \subset I$. Is there any element in the complement of J in I? If such an element exists, it might fix all the ambiguities of $\mathsf{F}_g^{\mathrm{alg}}$. This may be considered as a missing holomorphic anomaly equation.

For any regular function $f \in \mathscr{O}[\mathsf{T}]$ with $\mathbb{Q}$-coefficients and 8 elements $d_i \in \mathbb{Q}[\mathsf{R}_i,\ i = 0,1,\ldots,6]$ there is a polynomial P in 8 variables and with $\mathbb{Q}$-coefficients such that

$$P(d_1 f, d_2 f, \ldots, d_8 f) = 0.$$

The expression for P are usually big, see for instance [LY96a], and this may not give us a hint how to solve the ambiguities of $\mathsf{F}_g^{\text{alg}}$.

6.7 Topological partition functions $\mathsf{F}_g^{\text{alg}}, \ g=2,3$

Using Yamaguchi-Yau's computations and the expression for propagators in terms of t_i's, we can write $\mathsf{F}_2^{\text{alg}}$ [Supp Item 26] and $\mathsf{F}_3^{\text{alg}}$ [Supp Item 27] in terms of t_i's and we can check the corresponding anomaly equations:

$$\mathsf{R}_4\mathsf{F}_2^{\text{alg}} = \frac{1}{2}(\mathsf{R}_0^2\mathsf{F}_1^{\text{alg}} + (\mathsf{R}_0\mathsf{F}_1^{\text{alg}})^2)$$

for $g=2$ [Supp Item 28],

$$\mathsf{R}_4\mathsf{F}_3^{\text{alg}} = \frac{1}{2}\left(\mathsf{R}_0^2\mathsf{F}_2^{\text{alg}} + 2(\mathsf{R}_0\mathsf{F}_1^{\text{alg}})(\mathsf{R}_0\mathsf{F}_2^{\text{alg}})\right)$$

for $g=3$ [Supp Item 29].

6.8 Topological partition functions for elliptic curves

The reader may wonder why we have not yet discussed the topological partition functions F_g in the case of elliptic curves. The main reason is the absence of anomaly equations for such objects in the literature. Actually, in this case we have an explicit formula for F_g due to Dougals, Dijkgraaf, Kaneko and Zagier. In this section we explain this formula and leave it as an open problem to formulate the corresponding BCOV anomaly equation. This must be obtained using the $\mathfrak{sl}_2$ Lie algebra discussed in §2.12.

Let E be a complex elliptic curve and let $p_1,\dots,p_{2g-2}$ be distinct points of E, where $g>1$. We will discuss the case $g=1$ separately. The set $X_g(d)$ of equivalence classes of holomorphic maps $\phi: C\to E$ of degree d from compact connected smooth complex curves C to E, which have only one double ramification point over each point $p_i\in E$ and no other ramification points, is finite. By the Hurwitz formula the genus of C is equal to g. Define

$$N_{g,d} := \sum_{[\phi]\in X_g(d)} \frac{1}{|\operatorname{Aut}(\phi)|} \tag{6.10}$$

and

$$\mathsf{F}_g := \sum_{d=1}^{\infty} N_{g,d}q^d.$$

Mirror symmetry for elliptic curves, after Douglas and Dijkgraaf [Dij95], gives us the following description of F_g's: the generalized Jacobi theta function

$$\Theta(X,q,l) = \prod_{n>0}(1-q^n) \prod_{n>0,\ n \text{ odd}} (1-e^{n^2X/8}q^{n/2}l)(1-e^{-n^2X/8}q^{n/2}l^{-1}) \tag{6.11}$$

is considered as a formal power series in X and $q^{1/2}$ with coefficients in $\mathbb{Q}[l,l^{-1}]$. Let $\Theta_0(X,q)$ denote the coefficient of l^0 in Θ,

$$\Theta_0(X,q) = \sum_{n=0}^{\infty} A_n(q)X^{2n},$$

F_g is the coefficient of X^{2g-2} in $\log(\Theta_0)$. Recall that the classical Jacobi function is the following

$$\Theta(\tau|z) = \prod_{n=1}^{\infty}(1-q^{2\pi i\tau})(1+q^{2n-1}e^{2\pi iz})(1+q^{2n-1}e^{-2\pi iz}) = \sum_{n=-\infty}^{\infty} e^{\pi in^2\tau+2\pi inz} \tag{6.12}$$

where $q = e^{2\pi i\tau}$, see for instance [SS03]. It can be described using the geometry of elliptic curves in a similar way as Weierstrass $\wp$ function. In order to see the similarities between (6.12) and (6.11) we have to put $l = e^{2\pi iz}$. In this way τ, respectively z, parameterize the moduli of elliptic curves, respectively the elliptic curve itself. In (6.11) we have the second variable X, in which the series does not converge. After Zagier and Kaneko, see [KZ95], we know that

$$\mathsf{F}_g \in \mathbb{Q}[E_2,E_4,E_6].$$

For instance,

$$\mathsf{F}_2(q) = \frac{1}{103680}(10E_2^3 - 6E_2E_4 - 4E_6),$$

$$\mathsf{F}_3(q) = \frac{1}{35831808}(-6E_2^6+15E_2^4E_4-12E_2^2E_4^3+7E_4^3+4E_2^3E_6-12E_2E_4E_6+4E_6^2).$$

For $g=1$ we do note have ramification points and for $\phi : C \to E$ as before, $\mathrm{A}ut(\phi)$ consists of translations by elements of $\phi^{-1}(0)$ and so $\#\mathrm{A}ut(\phi) = d$. Therefore, $d \cdot N_{d,1} = \sum_{i|d} i$ is the number of group plus Riemann surface morphisms $C \to E$ of degree d. In this case we have the contribution of constant maps which is given by $N_{1,0}\log q = -\frac{1}{24}\log q$. Therefore,

$$q\frac{\partial \mathsf{F}_1}{\partial q} = -\frac{1}{24}E_2.$$

For more examples of F_g see the expository article [RY10]. Dijkgraaf in [Dij97] gives an anomaly equation whose solution after multiplication by an unknown quantity must be F_g. It is not clear what is such a quantity.

Chapter 7
Holomorphic differential Calabi-Yau modular forms

In this chapter we describe a field of meromorphic differential Calabi-Yau modular forms attached to a class of fourth-order linear differential equations which includes the Calabi-Yau equations in the sense of Almkvist-Enckevort-Straten-Zudilin. It is a differential field of transcendental degree 7 over the complex numbers and it can be considered as a generalization of the differential field generated by classical Eisenstein series of weight $2,4$ and 6, where the differential structure is given by the Ramanujan relations. The Bershadsky-Cecotti-Ooguri-Vafa anomaly equation can be solved using this field and so the higher genus partition functions are its elements. We also generalize the Yamaguchi-Yau generators for partition functions.

7.1 Fourth-order differential equations

In this section we list properties of a fourth-order differential equation used throughout the present chapter. We need these properties in order to solve the BCOV anomaly equation, do q-expansion and so on. At this stage we are not interested to obtain q-expansion with integral coefficients and so, the corresponding class of linear differential equations is bigger than the class of Calabi-Yau equations defined in [GAZ10].

Let us consider the fourth-order differential equation (2.41). We will assume that

1. We have

$$a_1 = -\frac{1}{2}a_2a_3 - \frac{1}{8}a_3^3 + \theta a_2 + \frac{3}{4}a_3\theta a_3 - \frac{1}{2}\theta^2 a_3. \tag{7.1}$$

This implies[1] that the monodromy group of L written in some basis is inside $\mathrm{Sp}(4,\mathbb{C})$ and hence it leaves a symplectic form invariant, see §7.4. The algebraic condition (7.1) can be written as

$$a_1 = -\frac{1}{2}a_3a_4 + \theta a_4. \tag{7.2}$$

[1] This is not equivalent, see Remark 7.

where a_4 is given in (2.42). We denote by $\langle \cdot, \cdot \rangle$ the corresponding non-degenerate, skew symmetric bilinear map in the solution space of $L=0$. This hypothesis produces polynomial relations between the solutions of $L=0$ and their derivatives, see Proposition 15.

2. We assume that $P_0(\theta) = \theta^2(\theta^2 + \cdots)$. This implies that there are two solutions of $L=0$ of the form

$$x_{21} := 1 + O(z),\ x_{11} := x_{21}\frac{\ln(z)}{2\pi i} + O(z).$$

We will need this hypothesis in order to do q-expansion at $z=0$, see §7.9. If $P_0(\theta) = \theta^4$ then $z=0$ is called the maximal unipotent monodromy singularity of $L=0$, MUM for short. All the Calabi-Yau equations in [GAZ10] have MUM singular points, but it was recently observed that some linear differential equations coming from the variation of Calabi-Yau varieties do not have such a singular point, see [CvS13].

3. We assume that $\langle x_{11}, x_{21} \rangle = 0$. We complete this into a basis x_{i1}, $i = 1,2,3,4$ such that $\langle x_{21}, x_{31} \rangle = 0$ and $\langle x_{11}, x_{31} \rangle = 1$, see Proposition 21 and Proposition 22. In summary, the intersection matrix in the basis x_{i1}, $i = 1,2,3,4$ is of the form

$$E := [\langle x_{i1}, x_{j1} \rangle] = \begin{pmatrix} 0 & 0 & 1 & * \\ 0 & 0 & 0 & * \\ -1 & 0 & 0 & * \\ * & * & * & 0 \end{pmatrix}. \tag{7.3}$$

4. The monodromy group of L written in the basis x_{i1}, $i = 1,2,3,4$ is inside $\mathrm{GL}(4,\mathbb{R})$. We will need this hypothesis in order to get monodromy invariant non-holomorphic functions constructed from the solutions of $L=0$ and their derivatives, see the definition of G-matrix in (7.7). I do not know any set of conditions on a_i's which is equivalent to this property.

5. The differential Galois group of $L=0$ is $\mathrm{Sp}(4,\mathbb{C})$. This follows, for instance, from the hypothesis (1) together with a maximal unipotent monodromy around some singularity, see Proposition 20.

Any Calabi-Yau equation listed in [GAZ10] satisfies the above conditions. However, note that in order to solve BCOV holomorphic anomaly equation we do not need the integrality of coefficients in q-expansions.

7.2 Hypergeometric differential equations

The reader may follow the text with hypergeometric differential equations (2.43), and in particular those of the form

$$L = \theta^4 - z(\theta+\alpha_1)(\theta+\alpha_2)(\theta+1-\alpha_1)(\theta+1-\alpha_2), \tag{7.4}$$

$$0 < \alpha_1 \leq \alpha_2 \leq \frac{1}{2}.$$

There are exactly 14 such equations with integral q-expansions, see Appendix C. In the table below we have collected the corresponding (α_1, α_2)'s.

(α_1, α_2)	(d, k)	$3\alpha_2 - 1 - \alpha_1$
$(\frac{1}{5}, \frac{2}{5})$	$(5,5)$	$+$
$(\frac{1}{6}, \frac{1}{3})$	$(3,4)$	$-$
$(\frac{1}{8}, \frac{3}{8})$	$(2,4)$	$+$
$(\frac{1}{10}, \frac{3}{10})$	$(1,3)$	$-$
$(\frac{1}{4}, \frac{1}{3})$	$(6,5)$	$-$
$(\frac{1}{6}, \frac{1}{4})$	$(2,3)$	$-$
$(\frac{1}{12}, \frac{5}{12})$	$(1,4)$	$+$
$(\frac{1}{4}, \frac{1}{2})$	$(8,6)$	$+$
$(\frac{1}{3}, \frac{1}{2})$	$(12,7)$	$+$
$(\frac{1}{6}, \frac{1}{2})$	$(4,5)$	$+$
$(\frac{1}{3}, \frac{1}{3})$	$(9,6)$	$-$
$(\frac{1}{4}, \frac{1}{4})$	$(4,4)$	$-$
$(\frac{1}{6}, \frac{1}{6})$	$(1,2)$	$-$
$(\frac{1}{2}, \frac{1}{2})$	$(16,8)$	$+$

Data of 14 hypergeometric equations

The monodromies around $z = 0$ and $z = 1$ written in a symplectic basis of the solution space of (7.4) are respectively

$$M_0 := \begin{pmatrix} 1 & 1 & 0 & 0 \\ 0 & 1 & 0 & 0 \\ d & d & 1 & 0 \\ 0 & -k & -1 & 1 \end{pmatrix} \quad M_1 := \begin{pmatrix} 1 & 0 & 0 & 0 \\ 0 & 1 & 0 & 1 \\ 0 & 0 & 1 & 0 \\ 0 & 0 & 0 & 1 \end{pmatrix},$$

where (d, k) are listed in the above table, see for instance [CYY08]. The partition $14 = 7 + 7$ according to whether $3\alpha_2 - 1 - \alpha_1 \geq 0$ or not appears in two different context. Brav and Thomas in [BT14] have shown that the first seven hypergeometric differential equations have monodromy group with infinite index in $\mathrm{Sp}(4, \mathbb{Z})$ and Singh and Venkataramana in [SV14, Sin14] have shown that the rest seven have finite index in $\mathrm{Sp}(4, \mathbb{Z})$. From a completely different point of view, this classification can be also obtained from the Kontsevich's formula for Lyapunov exponents, see Appendix D.

7.3 Picard-Fuchs equations

For Calabi-Yau equations coming from the variation of Calabi-Yau threefolds X_z with $\dim H^3(X_z, \mathbb{C}) = 4$, the functions x_{i1} can be written as periods of a holomorphic differential 3-form η over topological 3-cycles with complex coefficients:

$$\int_{\delta_i} \eta := x_{i1}, \ \delta_i \in H_3(X_z, \mathbb{C}).$$

In this way $\langle \cdot, \cdot \rangle$ corresponds to the intersection form in homology, that is,

$$\langle \int_{\delta_{z,1}} \eta, \int_{\delta_{z,2}} \eta \rangle := \langle \delta_{z,1}, \delta_{z,2} \rangle.$$

Note that for the purpose of this text we do not assume that $\delta_i \in H_3(X_z, \mathbb{Z})$. In this geometric context we have the algebraic de Rham cohomology $H^3_{\mathrm{dR}}(X_z)$ which is a $\mathbb{C}$-vector space of dimension 4 and has a canonical basis $\theta^i \eta := \nabla^i_\theta \eta$, $i = 0, 1, 2, 3$. Here ∇ is the Gauss-Manin connection of the family X_z. The linear differential equation L can be considered as a linear dependence of $\theta^4 \eta$ with respect to the elements of the basis:

$$\theta^4 \eta = \sum_{i=0}^{3} a_i(z) \theta^i \eta, \ \text{ in } H^3_{\mathrm{dR}}(X_z).$$

In $H^3_{\mathrm{dR}}(X_z)$ we have the pairing

$$H^3_{\mathrm{dR}}(X_z) \times H^3_{\mathrm{dR}}(X_z) \to \mathbb{C}, (\omega_1, \omega_2) \mapsto \int_{X_z} \omega_1 \cup \omega_2$$

which is Poincaré dual to the intersection from in the homology and we simply denote it again by $\langle \cdot, \cdot \rangle$. From now on we will use the notation of the geometric context, such as η, freely. However, note that every thing in this chapter can be formulated without the geometric considerations. Later in §9, we will use the Euler number χ and $(2,1)$ Hodge number h^{21} of X_z. We will set $h^{21} = 1$, assuming that X_z is a Calabi-Yau threefold whose third cohomology has Hodge numbers $1, 1, 1, 1$.

Remark 7 Any algebraic family of genus two curves $X_z, z \in \mathbb{P}^1$ with differential 1-forms ω_z of the second type (no residues around the poles) on X_z gives us a fourth-order Picard-Fuchs equation L satisfied by the periods of ω_z. In general, the condition (7.1) may not be valid for this L. For instance, for the hyperelliptic family, $y^2 - 4x^5 + \frac{3125}{256} z(x+1) = 0$, the periods of $\omega_z = \frac{dx}{y}$ satisfy the Picard-Fuchs equation

$$\frac{-590625z + 2304}{2560000z - 2560000} + \frac{-93z}{64z - 64}\theta + \frac{-515z - 16}{160z - 160}\theta^2 + \frac{-3z}{z-1}\theta^3 - \theta^4 = 0,$$

see for instance [Mov11b] for algorithms which compute such equations. The corresponding a_i's does not satisfy (7.1). For the computer codes used in this computation, see the author's webpage [Supp Item 30]. [2] We may try to find families of genus two hyperelliptic curves which satisfy our conditions on L. If such a family

[2] Note that for this example solutions of the indicial equation at $z = 0$, resp. $z = \infty$, are $\pm\frac{1}{10}, \pm\frac{3}{10}$, resp. $\frac{3}{8}, \frac{5}{8}, \frac{7}{8}, \frac{9}{8}$. Using $z^{\frac{1}{2}} \omega_z$, we get a linear differential equation whose indicial equations at $z = 0$ has solutions $\frac{i}{5}$, $i = 1, 2, 3, 4$.

exists, there must be a link between our theory of differential Calabi-Yau modular forms and Siegel modular forms.

From the Hodge theory point of view, the most relevant condition on L which has to do with a variation of Hodge structures of type $1,1,1,1$ and not $2,2$, is the condition (7.1). In Proposition 15 we have used this property to conclude that the intersection form $\langle,\rangle$ is compatible with the variation of Hodge structures of type $1,1,1,1$.

Sometimes it is useful to write $L(\eta)=0$ as a system:

$$\theta B = A\cdot B,$$

where

$$A=\begin{pmatrix} 0 & 1 & 0 & 0\\ 0 & 0 & 1 & 0\\ 0 & 0 & 0 & 1\\ a_0(z) & a_1(z) & a_2(z) & a_3(z)\end{pmatrix},\quad B:=\begin{pmatrix}\eta\\ \theta\eta\\ \theta^2\eta\\ \theta^3\eta\end{pmatrix}. \tag{7.5}$$

7.4 Intersection form

In a four-dimensional space there is only one non-degenerate, skew-symmetric bilinear map and so in the solution space of $L=0$ we can take a basis $x_{i1},\ i=1,2,3,4$ such that the intersection matrix $[\langle x_{i1},x_{j1}\rangle]$ is the 4×4 symplectic matrix Ψ. However, we will consider an arbitrary basis $x_{i1},\ i=1,2,3,4$ with the intersection matrix E. Surprisingly, it is not necessary to know the explicit format of E. In the geometric context, E is the intersection matrix in a basis in the homology and $E^{-\mathrm{tr}}$ is the intersection matrix in the Poincaré dual of such a basis. Let

$$x_{ij}:=\theta^{j-1}(x_{i1}),\ i,j=1,2,3,4, \tag{7.6}$$

$$\mathrm{X}=[x_{ij}].$$

Let us consider the intersection matrices

$$\mathrm{F}:=[\langle\theta^{i-1}\eta,\theta^{j-1}\eta\rangle]_{i,j=1,2,3,4}=\mathrm{X}^{\mathrm{tr}}E^{-\mathrm{tr}}\mathrm{X},$$

$$\mathrm{G}:=[\langle\theta^{i-1}\eta,\overline{\theta^{j-1}\eta}\rangle]_{i,j=1,2,3,4}=\mathrm{X}^{\mathrm{tr}}E^{-\mathrm{tr}}\overline{\mathrm{X}}. \tag{7.7}$$

Since the monodromy group is inside $\mathrm{Sp}(4,\mathbb{R})$, the matrix G is real matrix valued function on $\mathbb{P}^1-\mathrm{Sing}(L)$. Note that the matrices F and G do not depend on the choice of the basis $x_{i1},\ i=1,2,3,4$. Note also that

$$\mathrm{F}^{\mathrm{tr}}=-\mathrm{F},\ \overline{\mathrm{G}}^{\mathrm{tr}}=-\mathrm{G}.$$

We have the following differential equations for F and G:

$$\theta \mathrm{F} = A \cdot \mathrm{F} + \mathrm{F} \cdot A^{\mathrm{tr}}, \quad \overline{\theta}\mathrm{F} = 0, \tag{7.8}$$

and

$$\overline{\theta}\mathrm{G} = \mathrm{G} \cdot \overline{A}^{\mathrm{tr}}, \quad \theta \mathrm{G} = A \cdot \mathrm{G}, \tag{7.9}$$

where A is given in (7.5). It follows that

$$\theta \bar{\theta} \mathrm{G} = A \mathrm{G} \bar{A}^{\mathrm{tr}}.$$

The following trivial identities will be used frequently:

$$\theta \bar{\theta} |f|^2 = |\theta f|^2, \quad \theta \bar{\theta} \log |f| = 0,$$

$$\overline{\log f} = \log(\overline{f}), \quad f \text{ holomorphic}.$$

Remark 8 Let F be any matrix which satisfies (7.8). A simple calculation shows that $\theta(\mathrm{F}\mathrm{X}^{-1}) = A \cdot (\mathrm{F}\mathrm{X}^{-1})$. Since X is a fundamental system of solutions for $\theta \mathrm{X} = \mathrm{X} A^{\mathrm{tr}}$ we conclude that there is a constant matrix E such that $\mathrm{F} = \mathrm{X}^{\mathrm{tr}} E^{-\mathrm{tr}} \mathrm{X}$. This implies that there is a one to one correspondence between the intersection forms in the solution space of $L = 0$ (respectively monodromy invariant intersection forms) and the solutions of (7.8) (respectively solutions with rational functions).

Proposition 15 *There is an intersection form in the solution space of $L = 0$ such that*

$$\mathrm{F} := \mathrm{X}^{\mathrm{tr}} E^{-\mathrm{tr}} \mathrm{X} = a_5 \begin{pmatrix} 0 & 0 & 0 & 1 \\ 0 & 0 & -1 & -\frac{1}{2} a_3 \\ 0 & 1 & 0 & a_4 \\ -1 & \frac{1}{2} a_3 & -a_4 & 0 \end{pmatrix}, \tag{7.10}$$

where a_i's are given in (2.41) and (2.42).

Note that we do not claim that for any intersection form in the solution space of $L = 0$ the matrix F is of the form above. This may not be true and the reason will be clear in the proof.

Proof. According to Remark 8, it is enough to prove that the matrix in (7.10) satisfies the differential equation (7.8). In fact, the differential equation is satisfied if and only if we define a_4 as in (2.42) and we have the equality (7.2). For the computer codes used in this computation see the author's webpage [Supp Item 31].

Remark 9 For differential equations coming from geometry as it is explained (7.3), the intersection form (which is induced by the intersection form in homology) defines a matrix F with $\mathrm{F}_{12} = \mathrm{F}_{13} = 0$. This follows from the Griffiths transversality and the fact that F is the intersection form matrix in cohomology with Hodge numbers $1, 1, 1, 1$. Note that this is not in general true for families of genus two curve, see Remark 7, however particular families may exists. We can show that the only solution to (7.8) with the entries $\mathrm{F}_{12} = \mathrm{F}_{14} = 0$ is the matrix (7.10). For simplicity we use the geometry notation and $' = \theta$. Let $\mathrm{F}_{ij} := \langle \theta^{i-1}\eta, \theta^{j-1}\eta \rangle$. We start from the trivial equality $F_{15} - 2F'_{14} + F''_{13} = 0$. From this and our assumption we get $F_{14} = a_5$. From $0 = F'_{13} = F_{23} + F_{14}$ we get $F_{23} = -a_5$. We derivate further $F_{23} = -F_{14}$ and get $F_{24} = -\frac{1}{2}a_3F_{14}$. After further derivation of this equality, we get the expression for F_{34}. The calculation of F_{14} appears in [CdlOGP91], p. 49.

Proposition 16 *We have*

$$G\bar{F}^{-\mathrm{tr}}G^{\mathrm{tr}} = F. \tag{7.11}$$

Proof. Let A be the 4×4 matrix given by

$$\overline{[\theta^{i-1}\eta]}_{4\times 1} = A \cdot [\theta^{i-1}\eta]_{4\times 1}.$$

From this we get two equalities

$$\overline{F} = AG$$

and

$$\overline{F} = AFA^{\mathrm{tr}}.$$

Note that $\bar{A} = A^{-1}$. Calculating A from the first equality and replacing it in the second yields the desired equality.

From the equality (7.11) we can derive the following equality:

Proposition 17 *We have*

$$\theta\bar{\theta}(\log\theta\bar{\theta}\log G_{11}) = \frac{|a_5|^2}{(G_{11}\theta\bar{\theta}\log G_{11})^2} - 2\theta\bar{\theta}\log G_{11}. \tag{7.12}$$

Proof. The proof follows from Proposition 7.11 and can be done by any software in commutative algebra with Gröbner basis algorithm. For the computer codes used in this computation see the author's webpage [Supp Item 32].

The equalities (7.10) and (7.11) are the origin to all "geometry relations" in Physics literature. For further consequences of the equalities (7.10) and (7.11) see §9.

7.5 Maximal unipotent monodromy

For differential equations $L = 0$ such that $z = 0$ has a maximal unipotent monodromy, that is $P_0(\theta) = \theta^4$, we have the Frobenius basis of the solution space $L = 0$:

$$\begin{aligned}
\psi_0 &= 1 + O(z),\\
\psi_1 &= y_0 \cdot \ln z + g_1,\\
\psi_2 &= \frac{1}{2}y_0 \cdot (\ln z)^2 + g_1 \cdot \ln z + g_2,\\
\psi_3 &= \frac{1}{6}y_0 \cdot (\ln z)^3 + \frac{1}{2}g_1 \cdot (\ln z)^2 + g_2 \ln z + g_3,
\end{aligned}$$

where $g_i = O(z),\ i = 1,2,3$. We define

$$x_{21} := (2\pi i)^3\psi_0,\ x_{11} := (2\pi i)^2\psi_1,\ x_{31} := (2\pi i)\psi_2,\ x_{41} := -\psi_3. \tag{7.13}$$

The reason for $2\pi i$ factors is justified by the fact that the anti-clockwise monodromy around $z = 0$ and in the basis $\Delta = [x_{11}, x_{21}, x_{31}, x_{41}]^{\mathrm{tr}}$ is give by the matrix

$$M_0 := \begin{pmatrix} 1 & 1 & 0 & 0 \\ 0 & 1 & 0 & 0 \\ 1 & \frac{1}{2} & 1 & 0 \\ -\frac{1}{2} & -\frac{1}{6} & -1 & 1 \end{pmatrix}.$$

It is also justified by

Proposition 18 *In the solution space of $L = 0$ we consider a bilinear form given by the matrix*

$$E := [\langle x_{i1}, x_{j1}\rangle] = \begin{pmatrix} 0 & 0 & 1 & 0 \\ 0 & 0 & 0 & 1 \\ -1 & 0 & 0 & 0 \\ 0 & -1 & 0 & 0 \end{pmatrix}$$

where $x_{i1}, i = 1,2,3,4$ is the Frobenius basis at $z = 0$ given in (7.13). For 14 hypergeometric Calabi-Yau equations, this form is invariant under monodromy, that is, any analytic continuation of Δ resulting in $M\Delta$ satisfies

$$MEM^{\mathrm{tr}} = E. \tag{7.14}$$

Proof. This is based on explicit calculation of the monodromy group in [vEvS06] and [CYY08]. Note that a general E with the property (7.14) is defined up to multiplication with a non-zero constant and (7.14) with $M = M_0$ determines E with $E_{13} = 1$, except its $(3,4)$-entry. For the computer codes used in this computation see the author's webpage [Supp Item 33].

7.6 The field of differential Calabi-Yau modular forms

In this section we define the differential field of modular type functions attached to the fourth-order linear differential equation described in §7.1. We also describe their q-expansion.

Let $\mathbb{C}(z)(\mathrm{X})$ be the field of rational functions in the entries x_{ij} of X (defined in (7.6)) and with coefficients in $\mathbb{C}(z)$. The Proposition 15 gives us 6 non-trivial relations among x_{ij}.

Proposition 19 *If the differential Galois group of $L = 0$ is $\mathrm{Sp}(4,\mathbb{C})$ then the transcendental degree of $\mathbb{C}(z)(\mathrm{X})$ over $\mathbb{C}(z)$ is 10 and so any equality between x_{ij}'s over $\mathbb{C}(z)$ follows from Proposition 15.*

Proof. Let $k_1 := \mathbb{C}(z)$ and $k_2 := k_1(\mathrm{X})$. The matrix X is the fundamental system of the linear differential equation (7.5) and by our hypothesis the dimension of the Galois group of k_1 over k_2 is 10. This is the transcendental degree of the field k_2 over k_1, see for instance [vdPS03].

The monodromy group of $L = 0$ is a subgroup of $\mathrm{Sp}(4,\mathbb{C})$ and consequently, the differential Galois group $G(k_2/k_1)$ is an algebraic subgroup of $\mathrm{Sp}(4,\mathbb{C})$. If it contains a maximal unipotent matrix, for instant if the monodromy around $z = 0$ is MUM, then by a result of Saxl and Seitz, see [SS97], we have $G(k_2/k_1) = \mathrm{Sp}(4,\mathbb{C})$.

We define the following seven expressions:

$$\begin{aligned} u_0 &:= z, \\ u_i &:= x_{2i},\ i = 1,2,3,4, \\ u_5 &:= x_{21}x_{12} - x_{11}x_{22}, \\ u_6 &:= x_{13}x_{21} - x_{11}x_{23}. \end{aligned}$$

Note that the u_i, $i = 1,2,3,4$, respectively u_5, u_6, differs from those in §2.15 by a factor of $(2\pi i)^3$, respectively $(2\pi i)^5$.

Definition 2 Any rational function in u_i, $i = 0,\dots,6$ is called a differential Calabi-Yau modular form. The set

$$\mathscr{O}_{\mathsf{T}}{}^{\mathrm{hol}} := \mathbb{C}(u_0,u_1,u_2,u_3,u_4,u_5,u_6)$$

is called the field of differential Calabi-Yau modular forms.

Proposition 20 *If the differential Galois group of* $L = 0$ *is* $\mathrm{Sp}(4,\mathbb{C})$ *then the functions* u_i, $i = 0,1,\dots,6$ *are algebraically independent over* $\mathbb{C}$.

Proof. By Proposition 19, we can take x_{ij} as variables and define the ideal $I \subset \mathbb{C}(z)[\mathrm{X}]$ given by the entries of the equality (7.10). The affine variety given by $I = 0$ over $\mathbb{C}(z)$ has dimension 10. Consider seven parameters s_i, $i = 0,1,\dots,6$ and define $J = I + \langle u_i - s_i,\ i = 0,1,2,\dots,6\rangle \subset \mathbb{C}(s)[\mathrm{X}]$. Using a software in commutative algebra, we can verify that the affine variety given by $J = 0$ over $\mathbb{C}(s)$ is of dimension 4 and so there cannot be a polynomial relation between u_i's. Any such relation would imply that the variety over $\mathbb{C}(s)$ given by J is empty. For the computer codes used in this computation see the author's webpage [Supp Item 34].

Remark 10 Let us assume that $z = 0$ is a MUM singularity and so $P_0(\theta) = \theta^4$. We take the Frobenius basis as in (7.13). Its subfield of $\mathbb{C}(z)(\mathrm{X})$ invariant under the maximal unipotent monodromy M_0 is generated by 8 elements. We have already introduced six generators $u_i, i = 0,1,2,3,5,6$ (recall that $u_0 = z$). The two others are

$$y_1 := \frac{x_{31}}{x_{21}} - \frac{1}{2}\left(\frac{x_{11}}{x_{21}}\right)^2,\quad y_2 := \frac{1}{3}\left(\frac{x_{11}}{x_{21}}\right)^3 - \frac{x_{11}}{x_{21}}\frac{x_{31}}{x_{21}} - \frac{x_{41}}{x_{21}}. \tag{7.15}$$

7.7 The derivation

We define

$$\tau_0 := \frac{x_{11}}{x_{21}}, \quad q = e^{2\pi i \tau_0}.$$

Note that $\tau = 2\pi i \tau_0$ is used in §2.15. We have

$$\dot{*} := \frac{\partial *}{\partial \tau_0} = 2\pi i q \frac{\partial *}{\partial q} = \frac{u_1^2}{u_5} \theta * .$$

The field $\mathcal{O}_{\mathsf{T}}{}^{\text{hol}}$ together with the above derivation becomes a differential field.

Proposition 21 *We have the following ODE in u_i variables:*

$$\begin{cases} \dot{u}_0 = \frac{u_1^2}{u_5} u_0, \\ \dot{u}_1 = \frac{u_1^2}{u_5} u_2, \\ \dot{u}_2 = \frac{u_1^2}{u_5} u_3, \\ \dot{u}_3 = \frac{u_1^2}{u_5} u_4, \\ \dot{u}_4 = \frac{u_1^2}{u_5} \sum_{i=0}^{3} a_i(u_0) u_{i+1}, \\ \dot{u}_5 = \frac{u_1^2}{u_5} u_6, \\ \dot{u}_6 = \frac{u_1^2}{u_5} \left(2\frac{u_2 u_6 - u_3 u_5}{u_1} + \frac{1}{2} a_3(u_0) u_6 + a_4(u_0) u_5 + E_{12} \cdot a_5(u_0) \right), \end{cases} \tag{7.16}$$

where E_{12} is the intersection number of x_{11} and x_{12}.

Proof. The first six equalities are trivial. For the last equality we have used (7.10). For the computer codes used in this computation see the author's webpage [Supp Item 34].

From now on we assume that $E_{12} = 0$ and so the a_5 term does not appear in the last line of (7.16). In this way we get (2.44). The ordinary differential equation (7.16) with $E_{12} = 0$ is a generalization of R_0 (2.9) in a context where geometry is missing. For $E_{12} = 1$ it is a generalization of $\check{\mathsf{R}}_0$ in (2.15).

7.8 Yukawa coupling

Let x_{i1}, $i = 1, 2, 3, 4$ be an arbitrary basis of $L = 0$. The Yukawa coupling or genus zero partition function is defined to be

$$\mathsf{Y} = \frac{\partial^2}{\partial \tau_0^2} \left(\frac{x_{31}}{x_{21}} \right) = \left(\frac{u_1^2}{u_5} \theta \right)^2 \frac{x_{31}}{x_{21}},$$

We would like this to be a differential Calabi-Yau modular form and this put a restriction on the intersection matrix E.

Proposition 22 *If* $E_{23} = E_{12} = 0$, *where E is the intersection matrix in the basis* x_{i1}, *then the Yukawa coupling* Y *is a differential Calabi-Yau modular form and in fact*

$$\mathsf{Y} = a_5 \frac{u_1^4}{u_5^3} E_{13}.$$

Proof. Our proof is computational and follows from the polynomial relations (7.10) between x_{ij}'s. For the computer codes used in this computation see the author's webpage [Supp Item 34].

Remark 11 Let us define the expressions y_1 and y_2 as in (7.15) (we do not assume MUM property). We have

$$\frac{\partial^2 y_1}{\partial \tau_0^2} = \mathsf{Y} - 1.$$

Further, if the intersection matrix E is the symplectic matrix then we have

$$\frac{\partial y_2}{\partial \tau_0} = -2y_1.$$

This follows from

$$\frac{\partial}{\partial \tau_0} \frac{x_{41}}{x_{21}} = \frac{x_{31}}{x_{21}} - \frac{x_{11}}{x_{21}} \frac{\partial}{\partial \tau_0} \frac{x_{31}}{x_{21}}$$

which has to do with Griffiths transversality relation in Hodge theory, see §4.2.

7.9 q-expansion

Let us write

$$P_0(\theta) = \theta^2(\theta^2 + B_1\theta + B_0),$$
$$P_1(\theta) = \sum_{i=0}^{4} C_i\theta^i,$$

where the polynomials $P_i(\theta)$ are used in the expression of L in (2.41). We have assumed that P_0 has a double root at $\theta = 0$ in order to introduce q-expansions. This implies that $L = 0$ has two solutions of the form

$$\psi_0 = 1 + b_1 z + \cdots, \quad \psi_1 = \psi_0 \ln(z) + b_2 z + \cdots.$$

Note that in §7.8 we needed that $E_{12} = 0$, that is, the intersection of ψ_0 and ψ_1 is zero. Substituting ψ_0 and ψ_1 in $L = 0$ we calculate b_1 and b_2:

$$b_1 = \frac{C_0}{1+B_1+B_0}, \quad b_2 = \frac{(1+B_1+B_0)C_1 - C_0(4+3B_1+2B_0)}{(1+B_1+B_0)^2}.$$

For calculation of b_2 we have used

$$\theta^n(\psi_0 \ln(z)) = \theta^n \psi_0 + n\theta^{n-1}\psi_0.$$

Next, we see that the

$$q = e^{\frac{\psi_1}{\psi_0}} = z + b_2 z^2 + \cdots,$$

$$\psi_0 \theta \psi_1 - \psi_1 \theta \psi_0 = 1 + (b_2 + 2b_1)z + \cdots.$$

From all these and $u_6 = \theta u_5$, we get the initial data (2.45) in §2.15. The ODE (2.44) turns out to be a recursion in the coefficients of u_i's and the initial values (2.45) are enough for the recursion. For the computer codes used in this computation see the author's webpage [Supp Item 13].

Remark 12 If for a differential equation $L = 0$, u_1 and u_5 as functions in z, and u_0 as a function in q have integral coefficients then all the quantities u_i as functions of both z and q have integral coefficients. Note that we have $u_0 := z$ and we just substitute u_0 (as a power series in q) in all $u_i(z)$ and get the q-expansion of all u_i's.

Remark 13 The d-th coefficient of the q-expansion of the Yukawa coupling is conjecturally the number of rational curves in the A-model Calabi-Yau threefold of mirror symmetry. It is actually defined up to multiplication by a constant. Therefore, the first coefficient of Y is an ambiguity. For the number one Calabi-Yau equation, the first coefficient is $\kappa := 5$ and this is the degree of the quintic. In general

$$\kappa = \int_M \omega^3$$

where ω is the Kähler class of the A-model Calabi-Yau threefold M, see for instance Appendix of [HKQ09].

Chapter 8
Non-holomorphic differential Calabi-Yau modular forms

The main ingredients of the BCOV anomaly equation are non-holomorphic functions and anti-holomorphic derivations. This forces us to define the field of non-holomorphic differential Calabi-Yau modular forms. The starting point for defining holomorphic differential Calabi-Yau modular forms is the differential equation $\theta \mathrm{X} = \mathrm{X} A^{\mathrm{tr}}$ and in a similar way we can define non-holomorphic differential Calabi-Yau modular forms starting from the differential equation

$$\theta \mathrm{G}^{\mathrm{tr}} = \mathrm{G}^{\mathrm{tr}} A^{\mathrm{tr}}. \tag{8.1}$$

8.1 The differential field

Let $\mathbb{C}(z,\bar{z})(\mathrm{G})$ be the field of rational functions in the entries G_{ij} of G and with coefficients in $\mathbb{C}(z,\bar{z})$. The equality (7.11) gives us 6 non-trivial relations among G_{ij}'s. These are, in fact, the only possible relations:

Proposition 23 *If the differential Galois group of $L = 0$ is* $\mathrm{Sp}(4,\mathbb{C})$ *then the transcendental degree of* $\mathbb{C}(z,\bar{z})(\mathrm{G})$ *over* $\mathbb{C}(z,\bar{z})$ *is* 10.

Proof. For an analytic function $f(z)$ defined in a neighborhood of an open set in $U \subset \mathbb{C}$ with the coordinate system z, we have the holomorphic function $F : U \times U \to \mathbb{C}$ defined in a unique way by the equality $F(z,\bar{z}) = f(z)$. This implies that we can think of $\bar{z}$ as an independent variable and so substituting $\bar{z}$ with a constant makes sense so that f becomes holomorphic in z. In this way the matrix G^{tr} with a fixed $\bar{z}$, is holomorphic in z and it is a fundamental system of solutions for $\theta Y = Y A^{\mathrm{tr}}$. Since G satisfies the equation (7.11), similar to Proposition 19, the statement follows by general arguments in differential Galois theory. Note that looking G^{tr} as a holomorphic matrix, the intersection form in the basis of $L = 0$ given by the first column of G^{tr} is $\bar{\mathrm{F}}$.

The field $\mathbb{C}(z,\bar{z})(\mathrm{G})$ together with θ and $\bar{\theta}$ is a differential field and it has non-trivial differential subfields as follows. We use (8.1) and (7.11) and conclude that the $\mathbb{C}(z)$-algebra generated by

$$\begin{aligned}
u_1 &:= \mathrm{G}_{11},\\
u_2 &:= \mathrm{G}_{21},\\
u_3 &:= \mathrm{G}_{31},\\
u_4 &:= \mathrm{G}_{41},\\
u_5 &:= \mathrm{G}_{11}\mathrm{G}_{22} - \mathrm{G}_{21}\mathrm{G}_{12},\\
u_6 &:= \mathrm{G}_{11}\mathrm{G}_{32} - \mathrm{G}_{12}\mathrm{G}_{31}
\end{aligned}$$

is invariant under the holomorphic derivation θ:

$$\begin{cases}
\theta u_1 = u_2,\\
\theta u_2 = u_3,\\
\theta u_3 = u_4,\\
\theta u_4 = \sum_{i=0}^{3} a_i(z) u_{i+1},\\
\theta u_5 = u_6,\\
\theta u_6 = 2\dfrac{u_2u_6 - u_3u_5}{u_1} + \dfrac{1}{2}a_3(z)u_6 + a_4(z)u_5.
\end{cases} \tag{8.2}$$

Definition 3 We denote by $\mathscr{O}_{\mathrm{T}}{}^{\mathrm{non}}$ the field over $\mathbb{C}(z)$ generated by u_i, $i = 1, 2, \ldots, 6$ and call it the field of non-holomorphic differential Calabi-Yau modular forms.

The reader has noticed that we have used the same letter u_i for both holomorphic and non-holomorphic differential Calabi-Yau modular form. The link between these two apparently different notions is given by the holomorphic limit. The meaning of u_i will be clear in the context and hopefully this will not produce any confusion. Combining the arguments in Proposition 20 and Proposition 8.2, we conclude that non-holomorphic u_i's are also algebraically independent over $\mathbb{C}$.

Remark 14 In a similar way we have $\bar{\theta}\mathrm{G} = \mathrm{G}\bar{A}^{\mathrm{tr}}$ and the $\mathbb{C}(\bar{z})$-algebra generated by

$$v_i := \mathrm{G}_{1i},\ i = 1,2,3,4,\quad v_5 := \mathrm{G}_{11}\mathrm{G}_{22} - \mathrm{G}_{21}\mathrm{G}_{12}, v_6 := \mathrm{G}_{11}\mathrm{G}_{23} - \mathrm{G}_{21}\mathrm{G}_{13}$$

is invariant under the anti-holomorphic derivation $\bar{\theta}$. We get (8.2) replacing a_i with $\overline{a_i}$ and u_i with v_i

$$\begin{cases}
\bar{\theta} v_1 = v_2,\\
\bar{\theta} v_2 = v_3,\\
\bar{\theta} v_3 = v_4,\\
\bar{\theta} v_4 = \sum_{i=0}^{3} \overline{a_i(z)} v_{i+1},\\
\bar{\theta} v_5 = v_6,\\
\bar{\theta} v_6 = 2\dfrac{v_2v_6 - v_3v_5}{v_1} + \tfrac{1}{2}\overline{a_3(z)}v_6 + \overline{a_4(z)}v_5.
\end{cases} \tag{8.3}$$

8.2 Anti-holomorphic derivation

In this section we translate the anti-holomorphic derivation $\bar{\theta}$ to a collection of holomorphic derivations. The differential field $(\mathscr{O}_{\mathsf{T}}{}^{\mathrm{non}}, \theta)$ is not closed under the anti-holomorphic derivation and we may try to solve some differential equations in $\mathscr{O}_{\mathsf{T}}{}^{\mathrm{non}}$ involving $\bar{\theta}$. Here, is one example coming from Physics.

Let us consider a differential equation of the form

$$\frac{\bar{\theta}X}{\bar{a}_5} = \alpha,$$

where α is a known non-holomorphic differential Calabi-Yau modular form and X is an unknown quantity which in principle we would like to find it inside $\mathscr{O}_{\mathsf{T}}{}^{\mathrm{non}}$. We write

$$\bar{\theta}X = \sum_{i=1}^{6} (\bar{\theta}u_i)\frac{\partial X}{\partial u_i} = \alpha \cdot \bar{a}_5.$$

We have

$$\begin{aligned}
\bar{\theta}u_1 &= \mathrm{G}_{12}, \\
\bar{\theta}u_2 &= \mathrm{G}_{22}, \\
\bar{\theta}u_3 &= \mathrm{G}_{32}, \\
\bar{\theta}u_4 &= \mathrm{G}_{42}, \\
\bar{\theta}u_5 &= \mathrm{G}_{11}\mathrm{G}_{23} - \mathrm{G}_{21}\mathrm{G}_{13}, \\
\bar{\theta}u_6 &= \mathrm{G}_{11}\mathrm{G}_{33} - \mathrm{G}_{31}\mathrm{G}_{13}.
\end{aligned}$$

Here, all $\frac{\partial X}{\partial u_i}$, $i = 1, 2, \dots, 6$ and α belong to $\mathscr{O}_{\mathsf{T}}{}^{\mathrm{non}}$ but $\bar{\theta}u_i$ may not belong. However we have the polynomial relations (7.11) and modulo these relations we get:

$$\alpha_1\mathrm{G}_{33} + \alpha_2\mathrm{G}_{43} + \alpha_3\mathrm{G}_{42} + \alpha_4 + \bar{a}_5(\alpha_5 - \alpha) = 0, \ \alpha_i \in \mathscr{O}_{\mathsf{T}}{}^{\mathrm{non}}.$$

Using Proposition 24 below we conclude that $\alpha_1 = \alpha_2 = \alpha_3 = \alpha_4 = \alpha_5 - \alpha = 0$. After rescaling α_i's we get the following differential equations for X:

$$\begin{aligned}
0 &= u_5\frac{\partial X}{\partial u_5} + u_6\frac{\partial X}{\partial u_6}, \\
0 &= u_1\frac{\partial X}{\partial u_1} + u_2\frac{\partial X}{\partial u_2} + u_3\frac{\partial X}{\partial u_3} + u_4\frac{\partial X}{\partial u_4}, \\
0 &= \frac{1}{\beta_0}(\beta_1\frac{\partial X}{\partial u_1} + \beta_2\frac{\partial X}{\partial u_2} + \beta_3\frac{\partial X}{\partial u_3}), \\
0 &= \frac{1}{\beta_4}(\beta_5\frac{\partial X}{\partial u_5} + \beta_6\frac{\partial X}{\partial u_6}) - \alpha,
\end{aligned}$$

where β_i's are explicit polynomials in u_i and z:

$$
\begin{aligned}
\beta_0 &:= \mathrm{F}_{14}u_1^2u_4,\\
\beta_1 &:= (\mathrm{F}_{24}u_1^2u_6 - \mathrm{F}_{34}u_1^2u_5 - \mathrm{F}_{14}u_1u_2u_6 + \mathrm{F}_{14}u_1u_3u_5),\\
\beta_2 &:= (\mathrm{F}_{24}u_1u_2u_6 - \mathrm{F}_{34}u_1u_2u_5 + \mathrm{F}_{14}u_1u_4u_5 - \mathrm{F}_{14}u_2^2u_6 + \mathrm{F}_{14}u_2u_3u_5),\\
\beta_3 &:= (\mathrm{F}_{24}u_1u_3u_6 - \mathrm{F}_{34}u_1u_3u_5 + \mathrm{F}_{14}u_1u_4u_6 - \mathrm{F}_{14}u_2u_3u_6 + \mathrm{F}_{14}u_3^2u_5),\\
\beta_5 &:= (\mathrm{F}_{24}u_1u_3 + \mathrm{F}_{14}u_1u_4 - \mathrm{F}_{14}u_2u_3),\\
\beta_6 &:= (\mathrm{F}_{34}u_1u_3 - \mathrm{F}_{14}u_3^2),\\
\beta_4 &:= \frac{\mathrm{F}_{24}u_1u_3u_6 - \mathrm{F}_{34}u_1u_3u_5 + \mathrm{F}_{14}u_1u_4u_6 - \mathrm{F}_{14}u_2u_3u_6 + \mathrm{F}_{14}u_3^2u_5}{-\mathrm{F}_{14}u_1^2}.
\end{aligned}
$$

Note that $\mathrm{F}_{14} = a_5$, $\mathrm{F}_{24} = -\frac{1}{2}a_3a_5$, $\mathrm{F}_{34} = a_4a_5$ and a_i's are given in (2.41) and (2.42). For the computer codes used in this computation see the author's webpage [Supp Item 35].

Proposition 24 *If the differential Galois group of* $\mathsf{L} = 0$ *is* $\mathrm{Sp}(4,\mathbb{C})$ *then the elements* $\mathrm{G}_{33}, \mathrm{G}_{43}, \mathrm{G}_{42}$ *and* u_i, $i = 1,2,\ldots,6$ *are algebraically independent over* $\mathbb{C}(z,\bar{z})$.

Proof. Using Proposition (8.2) and similar to the proof of Proposition 20, we can consider G_{ij} as variables and consider the ideal I given by (7.11). Note that in this case we introduce more three parameters corresponding to $\mathrm{G}_{33}, \mathrm{G}_{43}, \mathrm{G}_{42}$.

Remark 15 Let us define

$$
\begin{aligned}
\tilde{\mathsf{R}}_1 &= u_5\frac{\partial}{\partial u_5} + u_6\frac{\partial}{\partial u_6},\\
\tilde{\mathsf{R}}_2 &= u_1\frac{\partial}{\partial u_1} + u_2\frac{\partial}{\partial u_2} + u_3\frac{\partial}{\partial u_3} + u_4\frac{\partial}{\partial u_4},\\
\tilde{\mathsf{R}}_3 &= \frac{1}{\beta_0}(\beta_1\frac{\partial}{\partial u_1} + \beta_2\frac{\partial}{\partial u_2} + \beta_3\frac{\partial}{\partial u_3}),\\
\tilde{\mathsf{R}}_4 &= \frac{1}{\beta_4}(\beta_5\frac{\partial}{\partial u_5} + \beta_6\frac{\partial}{\partial u_6}).
\end{aligned}
$$

We have shown that the differential equation $\frac{\bar{\theta}X}{\bar{a}_5} = \alpha$ is equivalent to

$$\tilde{\mathsf{R}}_1(X) = \tilde{\mathsf{R}}_2(X) = \tilde{\mathsf{R}}_3(X) = \tilde{\mathsf{R}}_4(X) - \alpha = 0. \tag{8.4}$$

In a similar way the differential equation $\bar{\theta}X = \alpha$ is equivalent to

$$\tilde{\mathsf{R}}_1(X) = \tilde{\mathsf{R}}_2(X) = \tilde{\mathsf{R}}_3(X) - \alpha = \tilde{\mathsf{R}}_4(X) = 0. \tag{8.5}$$

8.3 A new basis

We write the anti-holomorphic derivation using the variables:

$$w_1 := \mathsf{F}_{14} = a_5, \quad w_i := \frac{u_i}{u_1}, \quad i = 2,3,4, \quad w_5 := \frac{u_6}{u_5}.$$

Let us now assume that X is a function of w_i's and hence automatically we have

$$\tilde{\mathsf{R}}_1(X) = 0, \quad \tilde{\mathsf{R}}_2(X) = 0.$$

We can write down the $\tilde{\mathsf{R}}_i$, $i = 3,4$ in the w_i's in the following way:

$$\begin{aligned}
\tilde{\mathsf{R}}_3 &:= \frac{1}{\beta_0 u_1}((u_1\beta_2 - u_2\beta_1)\frac{\partial}{\partial u_2} + (u_1\beta_3 - u_3\beta_1)\frac{\partial}{\partial u_3} + (-\beta_1 u_4)\frac{\partial}{\partial u_4}) \\
&= \frac{1}{\beta_0 u_1^2}((u_1\beta_2 - u_2\beta_1)\frac{\partial}{\partial w_2} + (u_1\beta_3 - u_3\beta_1)\frac{\partial}{\partial w_3} + (-\beta_1 u_4)\frac{\partial}{\partial w_4}) \\
&= \frac{1}{\beta_0 u_1^2}(\mathsf{F}_{14}u_1^2 u_4 u_5 \frac{\partial}{\partial w_2} + \mathsf{F}_{14}u_1^2 u_4 u_6 \frac{\partial}{\partial w_3} + (-\beta_1 u_4)\frac{\partial}{\partial w_4}) \\
&= \frac{u_5}{u_1^2}(\frac{\partial}{\partial w_2} + w_5\frac{\partial}{\partial w_3} - (\frac{\mathsf{F}_{24}}{\mathsf{F}_{14}}w_5 - \frac{\mathsf{F}_{34}}{\mathsf{F}_{14}} - w_2 w_5 + w_3)\frac{\partial}{\partial w_4}) \\
&= \frac{u_5}{u_1^2}(\frac{\partial}{\partial w_2} + w_5\frac{\partial}{\partial w_3} - (-\frac{1}{2}a_3 w_5 - a_4 - w_2 w_5 + w_3)\frac{\partial}{\partial w_4}).
\end{aligned}$$

In a similar way

$$\begin{aligned}
\tilde{\mathsf{R}}_4 &:= \frac{u_5\beta_6 - u_6\beta_5}{\beta_4 u_5^2}\frac{\partial}{\partial w_5} \\
&= \frac{\mathsf{F}_{14}u_1^2\beta_4}{\beta_4 u_5^2}\frac{\partial}{\partial w_5} = a_5\frac{u_1^2}{u_5^2}\frac{\partial}{\partial w_5}.
\end{aligned}$$

Note that $\frac{\partial}{\partial u_i} = \frac{1}{u_1}\frac{\partial}{\partial w_i}$. We have also

$$\begin{cases}
\theta w_1 = \frac{1}{2}a_3(z)w_1, \\
\theta w_2 = w_3 - w_2^2, \\
\theta w_3 = w_4 - w_2 w_3, \\
\theta w_4 = a_0(z) + a_1(z)w_2 + a_2(z)w_3 + a_3(z)w_4 - w_2 w_4, \\
\theta w_5 = 2(w_2 w_5 - w_3) + \frac{1}{2}a_3(z)w_5 + a_4(z) - w_5^2.
\end{cases} \tag{8.6}$$

For the first equality we have used $\theta \mathsf{F}_{14} = \mathsf{F}_{24} + a_3\mathsf{F}_{14}$. For others we have used (8.2).

8.4 Yamaguchi-Yau elements

The following elements are natural generalizations of the elements introduced by Yamaguchi and Yau in [YY04] in the context of mirror quintic case:

$$
\begin{aligned}
X &:= 1 - w_1,\\
U &:= w_2,\\
V_1 &:= \frac{\theta^2(\frac{\mathrm{G}_{12}}{\mathrm{G}_{11}})}{\theta(\frac{\mathrm{G}_{12}}{\mathrm{G}_{11}})} + 2\frac{\theta \mathrm{G}_{11}}{\mathrm{G}_{11}} = w_5,\\
V_2 &:= \frac{\theta^2 \mathrm{G}_{11}}{\mathrm{G}_{11}} - V_1 \frac{\theta \mathrm{G}_{11}}{\mathrm{G}_{11}} = w_3 - w_5 w_2,\\
V_3 &:= \frac{\theta^3 \mathrm{G}_{11}}{\mathrm{G}_{11}} - \frac{\theta \mathrm{G}_{11}}{\mathrm{G}_{11}}(-V_2 - \frac{1}{2}a_3 V_1 - a_4)\\
&= w_4 - w_2(-w_3 + w_2 w_5 - \frac{1}{2}a_3 w_5 - a_4).
\end{aligned}
$$

Knowing that

$$
\begin{aligned}
w_1 &= 1 - X,\\
w_2 &= U,\\
w_3 &= V_2 + V_1 U,\\
w_4 &= V_3 + U(-V_2 - \frac{1}{2}a_3 V_1 - a_4),\\
w_5 &= V_1,
\end{aligned}
$$

we have

$$
\begin{aligned}
\tilde{\mathsf{R}}_4 &= a_5 \frac{u_1^2}{u_5^2}\frac{\partial}{\partial w_5} = a_5 \frac{u_1^2}{u_5^2}\left(\frac{\partial}{\partial V_1} - U\frac{\partial}{\partial V_2} - U(U - \frac{1}{2}a_3)\frac{\partial}{\partial V_3}\right),\\
\tilde{\mathsf{R}}_3 &= \frac{u_5}{u_1^2}\frac{\partial}{\partial U} = \frac{u_5}{u_1^2}\left(\frac{\partial}{\partial w_2} + w_5\frac{\partial}{\partial w_3} + (-w_3 + w_2 w_5 - \frac{1}{2}a_3 w_5 - a_4)\frac{\partial}{\partial w_4}\right).
\end{aligned}
$$

In Yamaguchi-Yau variables we have

$$
\begin{aligned}
\theta X &= \frac{1}{2}a_3 \cdot (X - 1),\\
\theta U &= V_2 + V_1 U - U^2,\\
\theta V_1 &= -2V_2 + \frac{1}{2}a_3 V_1 + a_4 - V_1^2,\\
\theta V_2 &= -a_3 \cdot U V_1 - V_1 V_2 - 2a_4 \cdot U + V_3,\\
\theta V_3 &= -a_3 \cdot U^2 V_1 - 2a_4 \cdot U^2 + (a_2 - \frac{1}{4}a_3^2 + a_4 + \frac{1}{2}\theta a_3) \cdot U V_1 - 2a_3 \cdot U V_2\\
&\quad + \frac{1}{2}a_3 \cdot V_1 V_2 + V_2^2 + (a_1 - \frac{1}{2}a_3 a_4 + \theta a_4) \cdot U + (a_2 + a_4) \cdot V_2 + a_3 \cdot V_3 + a_0.
\end{aligned}
$$

Note that using formulas (2.42) and (7.1), the last line of the above vector field can be written as:

$$\theta V_3 = -a_3 \cdot U^2 V_1 - 2a_4 \cdot U^2 + 2a_2 \cdot UV_1 - 2a_3 \cdot UV_2 + \frac{1}{2} a_3 \cdot V_1 V_2 + V_2^2$$
$$+2a_1 \cdot U + (a_2 + a_4) \cdot V_2 + a_3 \cdot V_3 + a_0.$$

For the computer codes used in this computation see the author's webpage [Supp Item 36].

8.5 Hypergeometric cases

Now, let us consider the case in which all the quantities $a_i(z),\ \ i = 0,1,2,3$ can be written as polynomials in $w_1 = a_5$. By our assumption all the elements in the right hand side of the differential equation (8.6) are in $\mathbb{Q}[w_1, w_2, w_3, w_4, w_5]$. Since $a_3(z) = P(w_1)$ for some polynomial P in one variable and coefficients in $\mathbb{Q}$, we have $\theta a_3 = P'(w_1)\theta w_1 = P'(w_1)(\frac{1}{2}a_3(z)w_1)$ which is again a polynomial in w_1. Our assumption in this section is valid for hypergeometric linear differential equations which we discuss it below.

Let us consider the hypergeometric linear differential equation (2.43). We have

$$a_i(z) = r_i \frac{z}{1-z} = r_i(w_1 - 1),\ \ w_1 = a_5 = \frac{1}{1-z}.$$

The condition (7.1) in this case is equivalent to $r_1 = r_2 = r_3 = 0$ or $r_1 = 4,\ r_2 = 6,\ r_3 = 4$ or

$$r_3 = 2,\ r_1 = r_2 - 1. \tag{8.7}$$

The 14 class of hypergeometric differential equations which comes from the variation of Calabi-Yau varieties lie in the last class and so from now on we assume (8.7). The corresponding differential equation can be written in the following format

$$L = \theta^4 - z \cdot (\theta - c_1)(\theta - c_2)(\theta - 1 + c_1)(\theta - 1 + c_2) = 0 \tag{8.8}$$

for some $c_1, c_2 \in \mathbb{C}$. The ordinary differential equation (8.6) turns out to be

$$\begin{cases} \theta w_1 = w_1^2 - w_1, \\ \theta w_2 = w_3 - w_2^2, \\ \theta w_3 = w_4 - w_2 w_3, \\ \theta w_4 = r_0(w_1 - 1) + r_1(w_1 - 1)w_2 + r_2(w_1 - 1)w_3 + 2(w_1 - 1)w_4 - w_2 w_4, \\ \theta w_5 = 2(w_2 w_5 - w_3) + (w_1 - 1)w_5 + \big((w_1 - 1)^2 + r_2(w_1 - 1) - w_1^2 + w_1\big) - w_5^2. \end{cases} \tag{8.9}$$

Chapter 9
BCOV holomorphic anomaly equation

Following Physics literature, we define

$$K := -\log(\mathrm{G}_{11}), \quad \theta K = -\frac{u_2}{u_1} = -w_2 = -U, \tag{9.1}$$

$$G_{z\bar{z}} := -\theta\bar{\theta}\log \mathrm{G}_{11} = -\frac{\mathrm{G}_{11}\mathrm{G}_{22}-\mathrm{G}_{12}\mathrm{G}_{21}}{\mathrm{G}_{11}^2} = -\frac{u_5}{u_1^2},$$

$$C_{zzz} := \mathrm{F}_{14} = a_5,$$

$$\Gamma_{zz}^z := -\frac{\theta G_{z\bar{z}}}{G_{z\bar{z}}} = -(\frac{u_6}{u_5}-2\frac{u_2}{u_1}) = -(w_5-2w_2) = -(V_1-2U),$$

$$C_{\bar{z}}^{zz} := \frac{\overline{\mathrm{F}_{14}}}{(\mathrm{G}_{11}\theta\bar{\theta}\log \mathrm{G}_{11})^2} = \overline{a_5}\frac{u_1^2}{u_5^2},$$

K is called the Kähler potential, $G_{z\bar{z}}$ is called the metric of the moduli space and Γ_{zz}^z is called the Christoffel symbol. The quantities right after the definition := can be defined for an arbitrary local coordinate t in $\mathbb{P}^1$ instead of z. For instance, the transformation of θK and Γ_{zz}^z after coordinate change is

$$\partial_t K = \frac{1}{\theta t}\theta K = -\frac{u_2}{u_1}, \quad \Gamma_{tt}^t = \frac{\theta^2 t}{(\theta t)^2} + \frac{\Gamma_{zz}^z}{\theta t}. \tag{9.2}$$

9.1 Genus 1 topological partition function

The genus one topological partition function $\mathsf{F}_1^{\mathrm{non}}$ is a function in z such that satisfies the anomaly equation:

$$\bar{\theta}\theta \mathsf{F}_1^{\mathrm{non}} = \frac{1}{2}\frac{|a_5|^2}{(\mathrm{G}_{11}\theta\bar{\theta}\log \mathrm{G}_{11})^2} + (\frac{\chi}{24} + \frac{1+h^{21}}{2})\theta\bar{\theta}\log \mathrm{G}_{11}.$$

Here, χ and $h^{21} = 1$ are respectively the Euler number and Hodge number of the underlying Calabi-Yau threefolds, as it is explained in §7.3. Note that this is the B-model Calabi-Yau threefold X of mirror symmetry and in the literature one usually use the notations of the A-model $\check{X}$ which satisfies $\chi(\check{X}) = -\chi(X)$ and $h^{11}(\check{X}) = h^{21}(X)$. For mirror quintic Calabi-Yau threefolds (the number one Calabi-Yau equation) we have $\chi := 200$. It follows from the equality (7.12) that the anomaly equation is solved by

$$\mathsf{F}_1^{\mathrm{non}} = -\frac{1}{2}\log\left((\theta\bar{\theta}\log \mathrm{G}_{11})^{-1}(\mathrm{G}_{11})^{-3-h^{21}-\frac{\chi}{12}}|f|^2\right).$$

See for instance [BCOV93], §2. Here f is a holomorphic function in z and is the ambiguity which cannot be solved by the anomaly equation. Recall that for any holomorphic function f we have $\theta\bar{\theta}\log|f|^2 = 0$. It is taken in the form (2.48) and so together with $\frac{\chi}{24} + \frac{1+h^{21}}{2}$ we have #Sing(L) constants which cannot be read from $L = 0$.

Remark 16 The partition function $\mathsf{F}_1^{\mathrm{non}}$ does not depend on the choice of the holomorphic differential form η, that is, if we multiply η with a holomorphic function then the right hand side of the above equality does not change and hence $\mathsf{F}_1^{\mathrm{non}}$ does not change. In a similar way, it does not depend on the choice of the vector field θ. The anomaly equation using the vector field $g(z)\theta$, for some holomorphic function $f(z)$, is obtained from the anomaly equation using θ multiplied by $|g|^2$.

9.2 The covariant derivative

In physics literature, there is a special bi-grading $\deg(x) = (n,m)$ of period quantities x. It interprets x as a section of the line bundle $L^n \otimes (T^*)^m$, where T is the tangent space of the base space $\mathbb{P}^1$ of $L = 0$ (parametrized by z) and L is the line bundle on $\mathbb{P}^1$ given by the holomorphic 3-forms on X_z (recall the geometric notations in §7.3). In this section we define this bi-grading in our context and the corresponding covariant derivative. We will avoid referring to the geometry of Calabi-Yau varieties and we will use only the linear differential equation $L = 0$. The weights in (2.11) are translated to

$$\deg(u_0) = 0, \;\; \deg(u_i) = 3, \quad i = 0,1,2,3, \; \deg(u_5) = \deg(u_6) = 5.$$

This is not used in mathematical physics and hence in the present section.

For a moment consider the quantities $\theta^j a_i, \; i = 0,1,2,3,5, \quad j = 0,1,2,\ldots$ as variables. We also consider new variables

$$f_n, \; n = 1,2,3,\ldots.$$

For any natural number $g \in \mathbb{N}$ let us define R_g to be the polynomial $\mathbb{Q}$-algebra generated by

$$a_5, \theta^j a_i, \ \theta^j f_n, \quad i=0,1,2,3, \quad j=0,1,2,\ldots, \quad n=1,2,3,\ldots,g \tag{9.3}$$

(the variables f_g are supposed to be the ambiguities of the topological partition functions $\mathsf{F}_g^{\mathrm{non}}$, $g \geq 2$ and $\theta \mathsf{F}_1^{\mathrm{non}}$ that will be defined later). In R_g we consider the bi-grading

$$\deg(f_g) = (0, 3g-3), \ g \geq 2, \ \deg(f_1) = (0,1), \tag{9.4}$$
$$\deg(\theta^j a_i) = (0, 4-i+j), \ i=0,1,2,3, \ j=0,1,2,\ldots, \ \deg(a_5) = (2,3).$$

We extend the definition of deg to the ring $R_g[w_2,w_3,w_4,w_5]$:

$$\deg(w_i) = (0, i-1), \ i=2,3,4, \ \deg(w_5) = (0,1).$$

Note that only the degree of a_5 has non-zero first entry. The following proposition follows immediately:

Proposition 25 *We have:*

1. *θ sends an element of degree (n,m) to an element of degree $(n,m+1)$.*
2. *If for X and Y in $R_g[w_2,w_3,w_4,w_5]$ we have $\bar{\theta}X = \frac{u_5}{u_1^2}Y$ and* $\deg(X) = (n,m)$ *then* $\deg(Y) = (n, m-1)$.
3. *If for X and Y in $R_g[w_2,w_3,w_4,w_5]$ we have $\bar{\theta}X = \bar{a}_5 \frac{u_1^2}{u_5^2}Y$ and* $\deg(X) = (n,m)$ *then* $\deg(Y) = (n, m-1)$.

Proof. The first item follows from the differential equation (8.6). The second item follows from the translation of the antiholomorphic derivation into holomorphic derivations (8.5), the formulas in §8.3 and our degree conventions (9.4). [1] The proof of the third item is similar to the second one.

Later, we will see the degree of topological partitions functions and propagators:

$$\begin{aligned} \deg(\mathsf{F}_g^{\mathrm{non}}) &= (2-2g, 0), \\ \deg(S) &= (-2, 0), \\ \deg(S^z) &= (-2, -1), \\ \deg(S^{zz}) &= (-2, -2). \end{aligned}$$

The covariant derivation is defined to be

$$\begin{aligned} \Theta x &:= (\theta + m\Gamma_{zz}^z + n\theta K)x \\ &= \left(\theta - m \cdot \theta \log(\theta\bar{\theta}\log(\mathrm{G}_{11})) - n \cdot \theta \log \mathrm{G}_{11}\right)x \\ &= (\theta - m(w_5 - 2w_2) - n \cdot w_2)x \end{aligned}$$

where $\deg(x) = (n,m)$.

[1] We do not need to define the bi-grading deg in $\mathscr{O}_\mathsf{T}$, however, this observation leads us to define $\deg G_{z\bar{z}} = \deg(\frac{u_5}{u_1^2}) = (0,0)$.

9.3 Holomorphic anomaly equation

The topological partition functions $\mathsf{F}_g^{\mathrm{non}}, \; g = 2, \dots$ are functions in z which satisfy the BCOV anomaly equation:

$$\bar{\theta}\mathsf{F}_g^{\mathrm{non}} = \frac{1}{2}C_{\bar{z}}^{zz}\left(\Theta\Theta\mathsf{F}_{g-1}^{\mathrm{non}} + \sum_{r=1}^{g-1}\Theta\mathsf{F}_r^{\mathrm{non}}\cdot\Theta\mathsf{F}_{g-r}^{\mathrm{non}}\right) \tag{9.5}$$

where $C_{\bar{z}}^{zz}$ is given in (9.1). If $\mathsf{F}_1^{\mathrm{non}}, \mathsf{F}_2^{\mathrm{non}}, \dots, \mathsf{F}_{g-1}^{\mathrm{non}}$ are given then $\mathsf{F}_g^{\mathrm{non}}$ is determined up to addition of a holomorphic function f_g.

9.4 Master anomaly equation

Let

$$Z := \exp\sum_{g=1}^{\infty}\mathsf{F}_g^{\mathrm{non}}\lambda^{2g-2}.$$

We have

$$\begin{aligned}
\bar{\theta}Z &= Z\cdot(\sum_{g=1}^{\infty}\bar{\theta}\mathsf{F}_g^{\mathrm{non}}\lambda^{2g-2})\\
&= \frac{1}{2}C_{\bar{z}}^{zz}Z(\sum_{g=2}^{\infty}\sum_{r=1}^{g-1}\Theta\mathsf{F}_r^{\mathrm{non}}\Theta\mathsf{F}_{g-r}^{\mathrm{non}}\lambda^{2g-2} + \sum_{g=2}^{\infty}\Theta^2\mathsf{F}_{g-1}^{\mathrm{non}}\lambda^{2g-2}) + Z\bar{\theta}\mathsf{F}_1^{\mathrm{non}}\\
&= \frac{1}{2}C_{\bar{z}}^{zz}\lambda^2\cdot Z((\Theta\log Z)^2 + (\Theta^2\log Z)) + (\bar{\theta}\mathsf{F}_1^{\mathrm{non}})Z\\
&= \frac{1}{2}C_{\bar{z}}^{zz}\lambda^2\Theta^2 Z + (\bar{\theta}\mathsf{F}_1^{\mathrm{non}})Z.
\end{aligned}$$

We can interpret ΘZ as

$$\begin{aligned}
\Theta Z &= Z(\sum_{g=1}^{\infty}\Theta\mathsf{F}_g^{\mathrm{non}}\lambda^{2g-2})\\
&= Z(\sum_{g=1}^{\infty}(\theta\mathsf{F}_g^{\mathrm{non}} - (2g-2)\theta K\mathsf{F}_g^{\mathrm{non}})\lambda^{2g-2})\\
&= \theta Z - (\theta K)\lambda\partial_\lambda Z.
\end{aligned}$$

Therefore, Z is annihilated by

$$\bar{\theta} - \frac{1}{2}C_{\bar{z}}^{zz}\lambda^2\Theta^2 - \bar{\theta}\mathsf{F}_1^{\mathrm{non}}, \quad \text{where} \quad \Theta := \theta - (\theta K)\lambda\partial_\lambda.$$

We have to calculate Θ^2 because we have a weighted derivation.

9.5 Algebraic anomaly equation

Using the computations in §8.2 we can break the anomaly equation into many pieces, where we write the anti-holomorphic derivation $\bar{\theta}$ in terms of holomorphic derivations. The BCOV anomaly equation is reduced to the following differential equations:

$$\begin{aligned} W_1(\mathsf{F}_g^{\text{non}}) &= 0, \\ W_2(\mathsf{F}_g^{\text{non}}) &= (W_0-(4-2g)w_2-(w_5-2w_2))(W_0-(4-2g)w_2)\mathsf{F}_{g-1}^{\text{non}} \\ &\quad +\sum_{r=1}^{g-1}(W_0-(2-2r)w_2)\mathsf{F}_r^{\text{non}}\cdot(W_0-(2-2(g-r))w_2)\mathsf{F}_{g-r}^{\text{non}} \end{aligned} \tag{9.6}$$

where the vector fields W_0, W_1, W_2 are given by

$$\begin{aligned} W_0 :=& \frac{1}{2}a_3w_1\frac{\partial}{\partial w_1}+(w_3-w_2^2)\frac{\partial}{\partial w_2} \\ &+(w_4-w_2w_3)\frac{\partial}{\partial w_3}+(a_0+a_1w_2+a_2w_3+a_3w_4-w_2w_4)\frac{\partial}{\partial w_4} \\ &+(2(w_2w_5-w_3)+\frac{1}{2}a_3w_5+a_4-w_5^2)\frac{\partial}{\partial w_5}, \\ W_1 :=& \frac{\partial}{\partial w_2}+w_6\frac{\partial}{\partial w_3}-(-\frac{1}{2}a_3w_5-a_4-w_2w_5+w_3)\frac{\partial}{\partial w_4}, \\ W_2 :=& (2w_1)\frac{\partial}{\partial w_5}. \end{aligned} \tag{9.7}$$

Recall the bi-grading in §9.2. Let $\deg_2$ be the second coordinate of the bi-grading deg defined in §9.2. In the proposition below, by degree we mean $\deg_2$.

Proposition 26 *The anomaly equation (9.5) can be solved by*

$$\mathsf{F}_g^{\text{non}} = \frac{1}{w_1^{g-1}}(P_g+p_g), \quad g\geq 2, \tag{9.8}$$

where P_g is a homogeneous polynomial of degree $3g-3$ in $R_g[w_2,w_3,w_4,w_5]$ such that vanishes at $(w_2,w_3,w_4,w_5)=(0,0,0,0)$ and $f_g=\frac{p_g}{w_1^{g-1}}$ is the ambiguity of $\mathsf{F}_g^{\text{non}}$.

Proof. We prove this by induction on g. First note that

$$\theta\mathsf{F}_1^{\text{non}} = \Theta\mathsf{F}_1^{\text{non}} = \frac{-1}{2}(-w_5+(-1-h^{21}-\frac{\chi}{12})w_2+f_1).$$

From this and the expression of W_i's the proposition follows for $g=2$. Let us assume that the proposition for $\mathsf{F}_r^{\text{non}}$, $r<g$. We have

$$\theta(\frac{A}{w_1^r}) = (-r)(\frac{1}{2}a_3(z))\frac{A}{w_1^r} + \frac{\theta A}{w_1^r}$$

and so

$$\Theta(\frac{A}{w_1^r}) = \frac{1}{w_1^r}(\theta - m(w_5 - 2w_2)w_1^r - n \cdot w_2 \cdot w_1^r - \frac{r}{2}a_3)A, \tag{9.9}$$

where $\deg(\frac{A}{w_1^r}) = (n,m)$. Therefore θ, and hence the covariant derivative Θ, does not increase the pole order at w_1. Using (8.6) we also conclude that if $X \in \mathbb{Q}[w_1, \frac{1}{w_1}]$ with pole order a at w_1 then $\Theta^n X$ has also the same property. Therefore, the left hand side of the anomaly equation is of the from $\frac{A}{w_1^{g-2}}$, where A is a homogeneous polynomial of degree $3g-4$. Now the assertion follows from the explicit expressions of W_i's.

Since derivation of w_1 along W_1 and W_2 is zero, we can use (9.9) and write down the anomaly equation of P_g in (9.8):

$$\begin{aligned} W_1(P_g) &= 0, \\ \tilde{W}_2(P_g) &= \left(W_0 - (4-2g)\left(w_2 w_1^{g-2} - \frac{a_3}{4}\right) - (w_5 - 2w_2)w_1^{g-2}\right) \\ &\quad \cdot \left(W_0 - (4-2g)\left(w_2 \cdot w_1^{g-2} - \frac{a_3}{4}\right)\right) P_{g-1} \\ &\quad + \sum_{r=1}^{g-1} \left(W_0 - (2-2r)\left(w_2 w_1^{r-1} - \frac{a_3}{4}\right)\right) P_r \\ &\quad \cdot \left(W_0 - (2-2(g-r))\left(w_2 w_1^{g-r-1} - \frac{a_3}{4}\right)\right) P_{g-r} \end{aligned}$$

where $\tilde{W}_2 = 2\frac{\partial}{\partial w_5}$ and W_0, W_1 is given in (9.7). Note that w_1 in W_1 is multiplied with $\frac{1}{w_1^{g-1}}$ of $\mathsf{F}_g^{\mathrm{non}}$ and it is canceled from both side of the anomaly equation (9.6).

For hypergeometric linear differential equations (2.43) we get:

Proposition 27 *If the ambiguity f_g is in $w_1^{g-1}\mathbb{Q}[w_1]$, then we have*

$$w_1^{g-1}\mathsf{F}_g^{\mathrm{non}} \in \mathbb{Q}[w_1, w_2, w_3, w_4, w_5], \ g \geq 2.$$

9.6 Proof of Theorem 9

Using Yamaguchi-Yau variables, the holomorphic anomaly equation for Q_g in (2.49) is the following:

$$R_1(Q_g) = \left(R_0 - (4-2g)\left((1-X)^{g-2}U - \frac{1}{4}a_3\right) - (1-X)^{g-2}(V_1 - 2U)\right)$$
$$\cdot\left(R_0 - (4-2g)\left((1-X)^{g-2}U - \frac{1}{4}a_3\right)\right)Q_{g-1}$$
$$+\sum_{r=1}^{g-1}\left(R_0 - (2-2r)\left((1-X)^{r-2}U - \frac{1}{4}a_3\right)\right)Q_r$$
$$\cdot\left(R_0 - (2-2(g-r))\left((1-X)^{g-r-1}U - \frac{1}{4}a_3\right)\right)Q_{g-r},$$

where the vector fields R_0, R_1 are given by

$$\begin{aligned} R_0 &:= \frac{1}{2}a_3\cdot(X-1)\frac{\partial}{\partial X} + (V_2 + V_1U - U^2)\frac{\partial}{\partial U} \\ &+(-2V_2 + \frac{1}{2}a_3V_1 + a_4 - V_1^2)\frac{\partial}{\partial V_1} \\ &+(-a_3\cdot UV_1 - V_1V_2 - 2a_4\cdot U + V_3)\frac{\partial}{\partial V_2} \\ &+(-a_3\cdot U^2V_1 - 2a_4\cdot U^2 + 2a_2\cdot UV_1 \\ &-2a_3\cdot UV_2 + \frac{1}{2}a_3\cdot V_1V_2 + V_2^2 + 2a_1\cdot U \\ &+(a_2+a_4)\cdot V_2 + a_3\cdot V_3 + a_0)\frac{\partial}{\partial V_3}, \\ R_1 &:= 2\left(\frac{\partial}{\partial V_1} - U\frac{\partial}{\partial V_2} - U(U - \frac{1}{2}a_3)\frac{\partial}{\partial V_3}\right). \end{aligned} \tag{9.10}$$

We know also that Q_g does not depend on U. The proof of Theorem 9 is similar to the proof of Proposition 9.8.

9.7 A kind of Gauss-Manin connection

Consider the case in which the linear differential equation L is the Picard-Fuchs equation of a family of algebraic varieties in the sense of §7.3, for instance take L the Picard-Fuchs equation of mirror quintic Calabi-Yau threefolds. In this case we can introduce a moduli space T of dimension 7 which is an affine variety and topological partition functions are certain regular functions on T, see §3. In this section we explain how this can be done in the absence of a geometry for L. The holomorphic anomaly equation (9.5) is written using θ and η which are global sections of the tangent bundle of $\mathbb{P}^1$ and the line bundle on $\mathbb{P}^1$ given by holomorphic 3-forms, respectively. We want to write it as an equation in a ghost moduli space T and using vector fields in T.

In the geometric context u_i's are interpreted as meromorphic functions on a moduli space T. They form a coordinates system for some Zariski open subset of T. Let X be the 4×4 matrix as in (7.6).

Proposition 28 *If the intersection matrix E in the basis x_{i1} is* $\begin{pmatrix} 0 & I_2 \\ -I_2 & 0 \end{pmatrix}$ *then there is a unique upper triangular matrix g such that $X\cdot g$ is of the τ-format* (4.5). *In fact*

$$g=\begin{pmatrix} \frac{1}{u_1} & -\frac{u_2}{u_5} & -\frac{u_2u_6-u_3u_5}{\mathsf{F}_{14}u_1^2} & -\frac{\mathsf{F}_{14}u_4+\mathsf{F}_{24}u_3-\mathsf{F}_{34}u_2}{\mathsf{F}_{14}^2} \\ 0 & \frac{u_1}{u_5} & \frac{u_6}{\mathsf{F}_{14}u_1} & \frac{\mathsf{F}_{14}u_3-\mathsf{F}_{34}u_1}{\mathsf{F}_{14}^2} \\ 0 & 0 & -\frac{u_5}{\mathsf{F}_{14}u_1} & -\frac{\mathsf{F}_{14}u_2-\mathsf{F}_{24}u_1}{\mathsf{F}_{14}^2} \\ 0 & 0 & 0 & \frac{u_1}{\mathsf{F}_{14}} \end{pmatrix}.$$

Proof. The proof is a mere computation [Supp Item 37]. The intersection matrix E in the format (7.3) is not enough to prove the proposition.

In the geometric context, each point of T represent a pair $(X,\{\alpha_i, i=1,2,3,4\})$, see §3. In our abstract context, where X is missing, α_i's are given by $\alpha=g^{\mathrm{tr}}B$, where

$$\alpha=[\alpha_1,\alpha_2,\alpha_3,\alpha_4]^{\mathrm{tr}},\ B:=[\eta,\theta\eta,\theta^2\eta,\theta^3\eta]^{\mathrm{tr}}.$$

The main reason to define α_i's in this way is as follows. We think of X as the period matrix of B in some symplectic basis of $H_3(X,\mathbb{Z})$, see §7.3. In this way, using Proposition 28 the period matrix of α over the same symplectic basis is of the τ-format. Note that in the mirror quintic case when we interpret t_i's as periods then this property holds for α_i's.

We write the linear differential equation L as $\theta B=A\cdot B$, where A is given in (7.5), and then in the α-basis:

$$\nabla\alpha=\mathsf{A}\otimes\alpha,$$

where A is given in (2.46).

9.8 Seven vector fields

In §8.2 we have derived four vector fields on T using the anti-holomorphic derivation $\bar{\theta}$. Composing these vector fields we get a particular format of four times four matrices. The matrices $\mathsf{A}_{\tilde{\mathsf{R}}_i},\ i=1,2$ are constant matrices and in fact $\tilde{\mathsf{R}}_i=\mathsf{R}_i,\ i=1,2$. However, $\mathsf{A}_{\tilde{\mathsf{R}}_i},\ i=3,4$ are not constants. We have to do a proper modification of $\tilde{\mathsf{R}}_i$'s which leads us to

$$\mathsf{R}_3=\tilde{\mathsf{R}}_3-s_2\tilde{\mathsf{R}}_2,\ \ \mathsf{R}_4=\tilde{\mathsf{R}}_4-s_1\tilde{\mathsf{R}}_1$$

where s_1,s_2 are given by

$$s_1 = \frac{-\mathsf{F}_{14}u_2u_6 + \mathsf{F}_{14}u_3u_5 + \mathsf{F}_{24}u_1u_6 - \mathsf{F}_{34}u_1u_5}{\mathsf{F}_{14}u_1^2u_4},$$
$$s_2 = \frac{-\mathsf{F}_{14}^2u_1^3u_4 + \mathsf{F}_{14}^2u_1^2u_2u_3 - \mathsf{F}_{14}\mathsf{F}_{24}u_1^3u_3}{\mathsf{F}_{14}u_1u_4u_5u_6 - \mathsf{F}_{14}u_2u_3u_5u_6 + \mathsf{F}_{14}u_3^2u_5^2 + \mathsf{F}_{24}u_1u_3u_5u_6 - \mathsf{F}_{34}u_1u_3u_5^2}.$$

This gives us an indication that the vector fields $\mathsf{R}_i,\ i = 1,2,3,4$ are intrinsic to the moduli space T despite the fact that we computed them through transcendental aspects of periods and anti-holomorphic derivations. Looking in this way take the Lie brackets of $\mathsf{R}_i,\ i = 0,1,2,3,4$ and find R_5 and R_6 which gives us the Lie algebra in Table (2.37). In a similar way we have the equations for Yukawa coupling (6.4), for $\mathsf{F}_1^{\mathrm{alg}}$ (6.5) and the master anomaly equation (6.8) in the context of fourth-order differential equations. For the computer codes used in this computation see the author's webpage [Supp Item 38].

9.9 Comparison of algebraic and holomorphic anomaly equations

In this section we describe the behaviour of many quantities under a change of η and the coordinate system z of $\mathbb{P}^1$. This will complete the translation of the original holomorphic anomaly equation (9.5) into the algebraic one introduced in §2.16.

Recall the quantities (9.1). Let us write $\tilde{\eta} = f\eta$, where f is a holomorphic function. In the following the tilde quantity corresponds to $\tilde{\eta}$ and without tilde corresponds to η. We have

$$\tilde{K}_z = K_z - \theta \log f.$$

The Christoffel symbol remains invariant, that is, $\tilde{\Gamma}_{zz}^z = \Gamma_{zz}^z$. The partition function $\mathsf{F}_g^{\mathrm{non}}$ transforms as

$$\tilde{\mathsf{F}}_g^{\mathrm{non}} = f^{2-2g}\mathsf{F}_g^{\mathrm{non}}.$$

This follows by induction and the structure of the anomaly equation of $\mathsf{F}_g^{\mathrm{non}}$. Note that

$$\begin{aligned}
\Theta \mathsf{F}_r^{\mathrm{non}} &= \Theta g^{2-2r}\tilde{\mathsf{F}}_r^{\mathrm{non}} \\
&= \theta(g^{2-2r}\tilde{\mathsf{F}}_r^{\mathrm{non}}) + (2-2r)K_z g^{2-2r}\tilde{\mathsf{F}}_r^{\mathrm{non}} \\
&= g^{2-2r}(\theta \tilde{\mathsf{F}}_r^{\mathrm{non}} + (2-2r)\tilde{K}_z\tilde{\mathsf{F}}_r^{\mathrm{non}}) \\
&= g^{2-2r}\tilde{\Theta}\tilde{\mathsf{F}}_r^{\mathrm{non}},
\end{aligned}$$

where $g := f^{-1}$, and

$$\tilde{C}_{\bar{z}}^{zz} = f^{-2}C_{\bar{z}}^{zz}.$$

Let us now analyze the quantities (9.1) under the change $\tilde{\theta} = f\theta$. We have

$$\tilde{K}_z = fK_z,$$

$$\tilde{\Gamma}_{zz}^{z} = f(\Gamma_{zz}^{z} - \theta \log f).$$

We have also

$$\tilde{C}_{\bar{z}}^{zz} = \bar{f} f^{-2} C_{\bar{z}}^{zz}.$$

By induction the topological partition function $\mathsf{F}_g^{\text{non}}$ is invariant under this transformation, that is, $\tilde{\mathsf{F}}_g^{\text{non}} = \mathsf{F}_g^{\text{non}}$. Note that

$$\tilde{\Theta}\tilde{\Theta}\tilde{\mathsf{F}}_r^{\text{non}} = f^2 \Theta\Theta \mathsf{F}_r^{\text{non}}.$$

Note that in the holomorphic anomaly equation for quantities with tilde we cancel the term f^2 and we obtain the corresponding anomaly for quantities without tilde. Now we perform the particular changes

$$\tilde{\theta} = \frac{u_1^2}{u_5}\theta, \; \tilde{\eta} = u_1^{-1}\eta$$

and we get

$$\tilde{K}_z = K_z + \frac{u_2}{u_1},$$

$$\tilde{\Gamma}_{zz}^{z} = \Gamma_{zz}^{z} + \theta \log(\frac{u_5}{u_1}).$$

Both quantities after taking holomorphic limit (in the sense of [BCOV94]) are zero. We have also the following equality:

$$\widetilde{\frac{\bar{\theta}}{C_{\bar{z}}^{zz}}} = \frac{\bar{\theta}}{\bar{a}_5}.$$

9.10 Feynman rules

One of the interesting observations in [BCOV94] is the discovery of Feynman rule for the topological partition functions $\mathsf{F}_g^{\text{non}}$, $g \geq 2$ using the anomaly equation (9.5). In this section we explain this.

We introduce the propagators which are the basic ingredients of the Feynman rule for the topological partition functions. The propagators S, S^z, S^{zz} satisfy the differential equations

$$\begin{aligned}
\bar{\theta} S^{zz} &= -C_{\bar{z}}^{zz}, \\
\bar{\theta} S^{z} &= (-\theta\bar{\theta} \log \mathrm{G}_{11}) \cdot S^{zz}, \\
\bar{\theta} S &= (-\theta\bar{\theta} \log \mathrm{G}_{11}) \cdot S^{z}
\end{aligned}$$

where $C_{\bar{z}}^{zz}$ is defined in (9.1).[2] If we fix solutions of the above differential equations then general solutions are given by

$$S^{zz}+f_1,\ S^z+f_2-f_1\theta\log \mathrm{G}_{11},\ S+f_3-f_2\theta\log \mathrm{G}_{11}+\frac{1}{2}f_1(\theta\log \mathrm{G}_{11})^2,$$

where f_1,f_2,f_3 are three holomorphic functions in z. The solutions of these differential equations are given by

$$\begin{aligned}
S^{zz} &:= \frac{-1}{a_5}\left(\theta\log(\mathrm{G}_{11}^2\theta\bar{\theta}\log\mathrm{G}_{11})\right)=\frac{-1}{a_5}\frac{u_6}{u_5}=\frac{-w_5}{w_1},\\
S^{z} &:= \frac{1}{a_5}\left((\theta\log\mathrm{G}_{11})^2+\theta^2\log\mathrm{G}_{11}\right)=\frac{1}{a_5}\frac{u_3}{u_1}=\frac{w_3}{w_1},\\
S &:= H_1\cdot(-\theta\log\mathrm{G}_{11})+H_2=\frac{F_{14}u_1u_4-F_{14}u_2u_3+F_{24}u_1u_3-F_{34}u_1u_2}{2F_{14}^2u_1^2},\\
&= \frac{w_4-w_2w_3+(-\frac{1}{2}a_3)w_3-a_4w_2}{2w_1},
\end{aligned}$$

where

$$H_1:=S^z-\frac{1}{2}\Theta S^{zz}-\frac{1}{2}(S^{zz})^2a_5=\frac{1}{2}\frac{\mathrm{F}_{34}}{a_5^2}=\frac{1}{2}\frac{a_4}{a_5},$$

$$H_2:=\frac{1}{2}\Theta S^z+\frac{1}{2}S^{zz}S^z a_5=\frac{F_{14}u_1u_4-F_{14}u_2u_3+F_{24}u_1u_3}{2F_{14}^2u_1^2}$$

and a_4,a_5 are defined in (2.42). The first equalities in each line are taken directly from [BCOV94], page 391. The second equalities correspond to our differential Calabi-Yau modular form notations introduced in §8. The third equalities correspond to w_i notation introduced in §8.3. The fact that H_1 is holomorphic in z is observed in [BCOV94], page 391 but it is not explicitly calculated there. Note that the bi-grading of S,S^z,S^{zz} are given by

$$\deg(S^{zz})=(-2,-2),\ \deg(S^z)=(-2,-1),\ \deg(S)=(-2,0).$$

For the computer codes used in this computation see the author's webpage [Supp Item 39].

The Feynman rule for computing $\mathsf{F}_g^{\mathrm{non}}$ is as follows. Let G be a graph whose vertices are indexed by $\mathbb{N}\cup\{0\}$ and with three types of edges 1: solid edges 2: dashed edges 3: half dashed and half solid edges. We denote the value of a vertex with index $g\in\mathbb{N}\cup\{0\}$ and with n-solid edges and m dashed edges as

$$\lambda^{2g-2+n+m}C_{n,m}^g$$

[2] In order to keep the compatibility of our notation and those in [BCOV94, YY04], $C_{\bar{z}}^{zz}$ must be taken the same object in the mentioned references but with the minus sign. In this way the partition function $\mathsf{F}_1^{\mathrm{non}}$ and S^{zz} are compatible with those in the Physics literature.

where λ is a new parameter (which is called the topological string coupling constant in the Physics literature) and [3]

$$
\begin{aligned}
C^g_{n,m} &:= \frac{(2g-2+n+m-1)!}{(2g-2+n-1)!}\Theta^n \mathsf{F}^{\rm non}_g, \text{ if } \ 2g-2+n \geq 1, \ g \geq 1 \\
&:= \frac{(n+m-3)!}{(n-3)!}\Theta^{n-3} w_1, \qquad \text{if } \ g=0, n\geq 3 \\
&:= 0, \qquad \text{if } \ g=0, n=0,1,2 \\
&:= (m-1)! \cdot (-\frac{\chi}{24}-1), \qquad \text{if } \ g=1, \ n=0, \ m\geq 1 \\
&:= 0, \qquad \text{if } \ g=1, \ n=0, \ m=0.
\end{aligned}
$$

The solid edges have the weight $-S^{zz}$, the half-solid and half dashed edges have the weight $-S^z$ and the dashed edges have the weight $-2S$. The weight of a graph is the products of all weights associated to its vertices and edges. The Feynman rule for $\mathsf{F}^{\rm non}_g$ is given by

$$
\mathsf{F}^{\rm non}_g \lambda^{2g-2} = - \sum_{\text{connected graphs } D \text{ of order } \lambda^{2g-2}} \frac{1}{\#(\mathrm{sym}(D))} \mathrm{weight}(D).
$$

In the Feynman rule of $\mathsf{F}^{\rm non}_g$ the sum is taken over all connected diagrams with $g-1 = \sum_{i=1}^k (g_i-1)+e_1+e_2+e_3$. Here k is the number of vertices and $g_i \in \mathbb{N}_0$ is the number attached to each vertices, e_1, respectively e_2, e_3, is the number of solid lines, respectively half dashed and dashed lines.

Since $C^0_{n,m} = 0$ for $n = 0,1,2$, we conclude that a diagram which may contribute a non-zero term to $\mathsf{F}^{\rm non}_g$ and with k vertices with index $g = 0$ satisfies $3k \leq 2e_1 + e_2$ and so $k \leq 2g-2$. For a connected diagram the number of vertices with index $g = 1$ is bounded by $e_1+e_2+e_3-1$ which is less than $2g-3$. The conclusion is that the sum in the Feynman rule is finite.

9.11 Structure of the ambiguity

In the case of a hypergeometric differential equation, we use (5.24) in [BCOV94] and the definition of a canonical coordinate in [BCOV94], page 338, (2.45), and conclude that the ambiguity f_g of $\mathsf{F}^{\rm non}_g$ is a rational function in z which is holomorphic everywhere except $z = 1, \infty$. At $z = 1$, respectively $z = \infty$, it has a pole of order at most $2g-2$, respectively $g-1$. Note that in this case $a_5 = (z-1)^{-1}$. This implies that f_g is of the forms

$$
f_g = \sum_{i=1}^{2g-2} (z-1)^{-i} + \sum_{i=0}^{g-1} (z^{-1})^{-i}.
$$

[3] The formulas given in [BCOV94], page 387 do not cover the case $g = 1, n = 0, m \geq 2$.

Let $X = 1 - a_5$ and so $z = \frac{X}{X-1}$. We conclude that $(1-X)^{g-1} f_g$ is a polynomial of degree at most $3g-3$ in X.

Chapter 10
Calabi-Yau modular forms

The classical theory of modular forms tells us that a modular form is a holomorphic function in the upper half plane satisfying a functional equation and a growth condition. One may wish to introduce the algebra of differential Calabi-Yau modular forms in a similar way. In this chapter we aim to do this in the framework of mirror quintics, and in general Calabi-Yau threefolds with one-dimensional complex moduli and with third cohomology of dimension 4. Actually, we will argue that t_0 and t_4, either as functions on the moduli space S in §3.2 or as q-series, are the true generalizations of Eisenstein series E_4 and E_6 and we will recover them through functional equations. We first recall a modification of the classical case of elliptic curves.

10.1 Classical modular forms

We consider the moduli $\mathbb{H}$ of $w := (E, \{\delta_1, \delta_2\})$, where E is an elliptic curve and $\{\delta_1, \delta_2\}$ is a basis of $H_1(E,\mathbb{Z})$ with $\langle \delta_1, \delta_2 \rangle = -1$. We have a holomorphic map on $\mathbb{H}$ which is given by

$$w \mapsto \tau(w) := \frac{\int_{\delta_1} \omega}{\int_{\delta_2} \omega},$$

where ω is any holomorphic 1-form in E. Because of the quotient the above map does not depend on the choice of ω. In fact this turns out to be a bijection between $\mathbb{H}$ and the upper half plane. Its inverse is given by

$$\tau \mapsto (\mathbb{C}/(\mathbb{Z}+\mathbb{Z}\tau), \{\delta_1, \delta_2\}),$$

where τ is a point in the upper half plane and δ_i, $i = 1,2$ is the closed cycle obtained by identifying the end points of the vector $\tau \in \mathbb{C}$, respectively $1 \in \mathbb{C}$. We have a natural action of $\mathrm{SL}(2,\mathbb{Z})$ on $\mathbb{H}$ given by

$$(A, w) \mapsto A(w) := (E, \{a\delta_1 + b\delta_2, c\delta_1 + d\delta_2\}), \quad A = \begin{pmatrix} a & b \\ c & d \end{pmatrix}.$$

Translated into the τ-coordinate, this is exactly the Möbius transformation on the upper half plane. A full modular form $f : \mathbb{H} \to \mathbb{C}$ of weight k is a holomorphic function such that satisfies the functional equation

$$(c \cdot \tau(w) + d)^{-k} f(A(w)) = f(w), \ \forall A \in \mathrm{SL}(2, \mathbb{Z}), \ w \in \mathbb{H}$$

and it has finite growth at cusps, that is, for any sequence in $\mathbb{H}$ which converges in sectors (see Definition 4) to a cusp, the limit of f evaluated at the sequence exists. The automorphy factor $j(A, w) = (c \cdot \tau(w) + d)$ gives us the line bundle ω on $\mathrm{SL}(2, \mathbb{Z}) \backslash \mathbb{H}$ such that $\omega \otimes \omega$ is the canonical bundle, for more details see [Mov12b] Appendix A.

10.2 A general setting

Let $\mathbb{H}$ be a, not necessarily simply connected, Riemann surface and let Γ be a finitely generated subgroup of $\mathrm{Sp}(4, \mathbb{Z})$ which acts discretely on $\mathbb{H}$:

$$\Gamma \times \mathbb{H} \to \mathbb{H}, \ (A, w) \mapsto A(w),$$

and hence $\Gamma \backslash \mathbb{H}$ inherits the structure of a, not necessarily compact, Riemann surface. This action may have elliptic points, that is, a point $w \in \mathbb{H}$ such that its stabilizer is a finite cyclic subgroup of Γ. We assume that there is a compactification $\overline{\Gamma \backslash \mathbb{H}} := \Gamma \backslash \mathbb{H} \cup \{\infty_i, \ i = 0, 1, 2, \ldots, n-1\}$ of this Riemann surface by adding n points which we call them cusps. The genus zero case $\overline{\Gamma \backslash \mathbb{H}} \cong \mathbb{P}^1$ will be of the main interest.

Definition 4 We say that $w \in \mathbb{H}$ converges to a cusp in sectors if its image under the composition $\pi : \mathbb{H} \to \Gamma \backslash \mathbb{H} \hookrightarrow \overline{\Gamma \backslash \mathbb{H}}$ converges in a sector $U \subset \overline{\Gamma \backslash \mathbb{H}}$ to a cusp and moreover, w lies in a finite number of connected components of $\pi^{-1}(U)$.

We consider four meromorphic functions τ_i, $i = 0, 1, 2, 3$ on $\mathbb{H}$ which satisfy the differential relations:

$$-\tau_3 d\tau_0 + d\tau_1 = 0, \ \ -\tau_1 d\tau_0 + \tau_0 d\tau_1 + d\tau_2 = 0$$

which we call them the Griffiths transversality relations. We define the 4×4 matrix

$$\tau(w) := \begin{pmatrix} \tau_0 & 1 & 0 & 0 \\ 1 & 0 & 0 & 0 \\ \tau_1 & \tau_3 & 1 & 0 \\ \tau_2 & -\tau_0\tau_3 + \tau_1 & -\tau_0 & 1 \end{pmatrix} \tag{10.1}$$

and we assume that for all $A \in \Gamma$,

$$A \cdot \tau(w) = \tau(A(w)) \cdot \mathrm{g}(A, w) \tag{10.2}$$

for a unique element $\mathsf{g}(A,w) \in \mathsf{G}$, where G is the Borel group in §3.10. This is a collection of functional equations for τ_i's. For instance,

$$\tau_0(A(w)) = \frac{A_{11}\tau_0(w) + A_{12} + A_{13}\tau_1(w) + A_{14}\tau_2(w)}{A_{21}\tau_0(w) + A_{22} + A_{23}\tau_1(w) + A_{24}\tau_2(w)}.$$

Note that (10.2) implies that $\mathsf{g}(A,w)$ is an automorphy factor, that is,

$$\mathsf{g}(AB,w) = \mathsf{g}(A,B(w)) \cdot \mathsf{g}(B,w), \quad w \in \mathbb{H}, \ A,B \in \Gamma.$$

This implies that

$$j(A,w) := (A_{21}\tau_0(w) + A_{22} + A_{23}\tau_1(w) + A_{24}\tau_2(w)), \ w \in \mathbb{H}, \ A \in \Gamma$$

is also an automorphy factor. Note that the functions τ_i's are meromorphic and so $\mathsf{g}(A,w)$ and j are also meromorphic functions on $\mathbb{H}$.

Definition 5 A Calabi-Yau modular form of weight k is a holomorphic function $f : \mathbb{H} \to \mathbb{C}$ which satisfies:

1. Functional equation: for all $A \in \Gamma$,

$$j(A,w)^{-k} f(A(w)) = f(w).$$

2. Growth condition: For $w \in \mathbb{H}$ which converges to a cusp in sectors the limit $f(w)$ exists.

We may try to produce Calabi-Yau modular forms by getting convergent Poincaré series

$$\sum_{A \in \Gamma} j(A,w)^k f(A(w)),$$

where $f : \mathbb{H} \to \mathbb{C}$ is a holomorphic function. The first step would be to prove a convergence theorem for $\sum_{A\in\Gamma} |j(A,w)|^{-2}$. In the case of elliptic curves, the proof of convergence uses the equality $d(A\tau) = j(A,\tau)^{-2} d\tau$, see for instance Freitag's book [Fre83]. This property in our context does not exit. It may be also proved using the Poincaré metric of the upper half plane and the fact that its quotient by $\mathrm{SL}(2,\mathbb{Z})$ has finite volume. For this reason in Appendix B we have worked out the generalization of Poincaré metric in the case of mirror quintics. We are not going to analyze the algebra of Calabi-Yau modular forms in such a generality, instead we focus on the case of mirror quintics and prove that it is a polynomial ring in two variables.

10.3 The algebra of Calabi-Yau modular forms

Recall the notations $\tilde{\mathbb{H}}$, $\mathbb{H}$, Γ in §4.6 and the monodromy matrices M_0, M_1 in §2.7. From the results of Brav-Thomas [BT14] it follows that $\tilde{\mathbb{H}}$ has infinitely many components and the monodromy representation

$$\pi_1(\mathbb{P}^1\backslash\{0,1,\infty\})/\langle\gamma_\infty^5=1\rangle \to \mathrm{Sp}(4,\mathbb{Z})$$

is injective, here γ_∞ is a closed path in $\mathbb{P}^1$ which encircles the infinity. It follows also that $\mathbb{H}$ is biholomorphic to the upper half plane, however, for our discussion of Calabi-Yau modular forms we will never use the coordinate on $\mathbb{H}$ given by this biholomorphism. For $w\in\mathbb{H}$ we denote the corresponding pair by $(X(w),\delta(w))$. Let $z:\mathbb{H}\to\mathbb{C}$ be the meromorphic function on $\mathbb{H}$ defined by the equality $X(w)=\mathsf{X}_{1,z(w)}$, where $\mathsf{X}_{1,z}$ is the mirror quintic defined in §3.1. It has a pole of order 5 at the elliptic points corresponding to $\mathsf{X}_{1,\infty}$. Let $w_1\in\mathbb{H}$ be a point with $z(w_1)$ near to zero and such that the monodromy group written in $\delta(w_1)$ is given by the explicit matrices $M_0,M_1\in\mathrm{Sp}(4,\mathbb{Z})$ in (2.26) in the Introduction. For $w\in\mathbb{H}$ there is a path in $\mathbb{P}^1\backslash\{0,1,\infty\}$ which connects $z(w)$ to $z(w_1)$ and such that $\delta(w)$ is the monodromy of $\delta(w_1)$ along this path. Moreover, this path is unique up to homotopy and up to $\gamma_\infty^5=1$. We define

$$\tau_i:\mathbb{H}\to\mathbb{C},\ i=0,1,2,$$

$$\tau_0(w)=\frac{\int_{\delta_1}\omega}{\int_{\delta_2}\omega},\ \tau_1(w)=\frac{\int_{\delta_3}\omega}{\int_{\delta_2}\omega},\ \tau_2(w)=\frac{\int_{\delta_4}\omega}{\int_{\delta_2}\omega},$$

where ω is a holomorphic differential form on $X(w)$. They do not depend on the choice of ω. We can write τ_i as in (2.34). This example has two cusps $\infty_i,\ i=0,1$ with their corresponding monodromy matrices M_0 and M_1. The particular formats of $M_i,\ i=0,1$ imply that

$$\tau_0(M_0(w))=\tau_0(w)+1,\quad \tau_i(M_1(w))=\frac{\tau_i(w)}{1+\tau_2(w)}.$$

Recall the moduli space S for the mirror quintic Calabi-Yau threefolds in §3.2. For $\mathsf{t}\in\mathsf{S}$ we denote the corresponding pair by $(X(\mathsf{t}),\omega(\mathsf{t}))$. Recall also the variety $\mathsf{X}_{1,z(w)}$ and the differential 3-form $\eta(w):=\eta$ on it. We have two Calabi-Yau modular forms $s_0,\ s_4$ of weights $1,5$, respectively:

$$s_0(w):=\int_{\delta_2(w)}\eta(w), \tag{10.3}$$

$$s_4(w):=z(w)\Big(\int_{\delta_2(w)}\eta(w)\Big)^5.$$

Therefore, degree m polynomials in s_0,s_5 with weights $\mathrm{weight}(s_0)=1$, $\mathrm{weight}(s_4)=5$, are Calabi-Yau modular forms of weight m. We prove that these are all.

Theorem 12 *A Calabi-Yau modular form of weight m can be written as a homogeneous polynomial of degree m in s_0,s_4 with weights* $\mathrm{weight}(s_0)=1,\ \mathrm{weight}(s_4)=5$.

Proof. We have a canonical map

$$\mathbb{H}\to\mathsf{S},\ w\mapsto(X(w),\omega(w)),$$

where $\omega(w) := \frac{\eta(w)}{\int_{\delta_2(w)}\eta(w)}$ is the unique holomorphic 3-form in $X(w)$ such that

$$\int_{\delta_1(w)} \omega(w) = \tau_0(w), \quad \int_{\delta_2(w)} \omega(w) = 1.$$

The integral $\int_{\delta_2(w)} \eta(w)$ may have zeros and so the above map is meromorphic. It is not necessarily injective. For instance, w and $M_0(w)$ are mapped to the same point. A Calabi-Yau modular form f of weight k induces a holomorphic map $\hat{f} : \mathsf{S} \to \mathbb{C}$ defined uniquely through the properties:

1.
$$\hat{f}(t \bullet a) = a^{-k}\hat{f}(t), \; a \in \mathbb{G}_m, \; t \in \mathsf{S}. \tag{10.4}$$

2. The pull-back of $\hat{f}$ by $\mathbb{H} \to \mathsf{S}$ is f.

In fact $\hat{f}$ can be defined explicitly:

$$\hat{f} : \mathsf{S} \to \mathbb{C}, \quad \hat{f}(t) = f(w)\Big(\int_{\delta_2(w)} \omega(t)\Big)^{-k},$$

where for t we choose any $w \in \mathbb{H}$ such that $X(w) = X(t)$. The functional equation of f implies that $\hat{f}$ is a well-defined map: any other choice of w is $A(w)$ for some $A \in \Gamma$ and

$$f(A(w))\Big(\int_{\delta_2(A(w))} \omega(t)\Big)^{-k} = f(w)\Big(\int_{\delta_2(w)} \omega(t)\Big)^{-k}$$

which is the reformulation of the functional equation of f. If for a choice of w we have $\int_{\delta_2(w)} \omega(t) = 0$ then we can simply choose another $A(\delta)$, $A \in \Gamma$ such that the corresponding integral does not vanish. This implies that the zeros of $\int_{\delta_2(w)} \omega(t)$ do not produce poles for $\hat{f}$. It turns out that $\hat{s}_i = t_i, \;\; i = 0, 4$. We have an isomorphism $\mathsf{S} \cong \mathbb{C}^2 \backslash \{t_4(t_4 - t_0^5) = 0\}$ and the composition $H : \mathbb{H} \to \mathsf{S} \to \mathbb{C}^2$ is given by $(s_0, s_4) : \mathbb{H} \to \mathbb{C}^2$. We are going to analyze the image of $\mathbb{H}$ near the points w with $z(w)$ near to 0 and 1. Let $w_1 \in \mathbb{H}$ be as in the beginning of the present section. The integral $\int_{\delta_2(w_1)} \eta(w_1)$ is holomorphic at $z = 0$ and using its first and second coefficients we conclude that the image of H contains a disc which is transversal to $t_4 = 0$ at some point. The growth condition of f implies that f restricted to D is holomorphic. Using the property (10.4) we conclude that $\hat{f}$ extends holomorphically to $\{t_4 = 0\} \backslash \{t_4 - t_0^5 = 0\}$. Now, let us make the monodromy of w_1 to a point w with $z(w)$ near 1 and along the interval $(0, 1)$. Instead of a disc we get a spiral surface with a center p in $t_4 - t_0^5 = 0$. This follows from the Frobenius basis calculation in §4.18. The growth condition on f, implies that the limit of $\hat{f}$ restricted to this surface exists as far as we approach p in sectors in the spiral surface. Again, this together with the property (10.4) imply that $\hat{f}$ extends to a holomorphic function in $\{t_4 - t_0^5 = 0\} \backslash \{t_4 = 0\}$. Finally we get a homomorphic function in $\mathbb{C}^2$ with the property (10.4). This implies the desired statement.

One may also try to prove Theorem 12 in the following way: Let us take a modular form f of weight k. The quotient $\frac{f}{s_0^m}$ is invariant under the group Γ and so it induces a meromorphic function on $\Gamma\backslash\mathbb{H}$ which we denote it by g. Since the limits of g in sectors around the cusps exist, we conclude that g is a meromorphic function in the compactification $\mathbb{P}^1$ of $\Gamma\backslash\mathbb{H}$. The poles of g arise from the zeros of t_0. We already know that the elliptic point $e \in \mathbb{H}$ is a zero of order 1. However, we do not have any information about other zeros of t_0. They may exists and so, they produce difficulties for identifying f in terms of t_0, t_4.

Chapter 11
Problems

Topological string theory and in particular mirror symmetry have provided us with many generating functions which are similar with, but different from, the classical modular forms. In the previous sections we gave a candidate for what would be the theory of modular forms in the case of mirror quintics. We called them differential Calabi-Yau modular forms. Since we have a wide variety of arithmetic applications for modular forms, it is desirable to see similar applications for differential Calabi-Yau modular forms. In our way, we encounter many problems and conjectures, which are not the main focus in the Physics literature. The main objective of this chapter is to fill this gap and give a collection of problems and conjectures whose solutions may lead to a better understanding of differential Calabi-Yau modular forms and in particular topological string partition function. Almost all the problem that we discuss in the present text are open for mirror quintics and so we focus our attention on this example. The formulation of such problems to other Calabi-Yau varieties/equations is not a difficult task and is left to the reader.

11.1 Vanishing of periods

Our first problem is:

Problem 1 *Let X be a mirror quintic and η be a holomorphic 3-form on X. Determine all or some of X and $\delta \in H_3(X,\mathbb{Z})$ such that the integral $\int_\delta \eta$ vanishes.*

Let us consider the family X_z, $z \in \mathbb{P}^1$ of mirror quintics. By definition we have

$$\int_{H_3(X_z,\mathbb{Z})} \eta = \mathbb{Z}[x_{11}, x_{21}, x_{31}, x_{41}]$$

where x_{ij} are defined as in Theorem 4. Therefore, our problem is reduced to study the zeros of the analytic continuations of $f(z) = n_1x_{11} + n_2x_{21} + n_3x_{31} + n_4x_{41}$, $n_i \in \mathbb{Z}$ in $\mathbb{P}^1 - \{0, 1, \infty\}$. We may analyze such zeros using a computer as follows.

We consider any uniformization $\alpha : \mathbb{H} \to \mathbb{P}^1 - \{0,1,\infty\}$ with ramification index $\infty, \infty, 5$ at $0,1,\infty \in \mathbb{P}^1$, respectively. We may further assume that $\alpha(\tau+1) = \alpha(\tau)$, $\alpha(\tau) = \sum_{n=0}^{\infty} \alpha_i e^{2\pi i \tau}$ and α maps $i\infty$ to zero. This series converges in the unit disc. For a multivalued holomorphic function f in $\mathbb{P}^1 - \{0,1,\infty\}$ with the above ramification index, the composition $f(\alpha(\tau))$ is one valued in the unit disc. If it has zeros then its inverse has convergence radius strictly less than one which can be approximated by Hadamard formula. Kontsevich in a talk in 2013, see [Kon13] and [Supp Item 40], claimed that he has done this procedure and he has found zeros of the holomorphic solution x_{21}. Note that a zero of x_{21} would give us an essential singularity for the mirror map $q = e^{2\pi i \frac{x_{11}}{x_{21}}}$, where $x_{11} = x_{21} \frac{\ln(z)}{2\pi i} + O(z)$ is the logarithmic solution. He also claimed that using the above method he has done many numerical computations in support of the following:

Problem 2 *Let ψ_0 and ψ_1 be the holomorphic and logarithmic solutions of the linear differential equation (2.25) at $z = 0$. The analytic continuations of the Wronskian $\psi_0 \theta \psi_1 - \psi_1 \theta \psi_0$ in $\mathbb{P}^1 - \{0,1,\infty\}$ has no zero.*

Note that $\psi_0 \theta \psi_1 - \psi_1 \theta \psi_0$ up to a power of $2\pi i$ factor is $x_{11}x_{22} - x_{12}x_{22}$. Kontsevich needed to verify Problem 2 in his study of Lyapunov exponents for the variation of Hodge structures over the punctured line and arising from mirror quintics, see Appendix D.

The zeros of x_{21} give us zeros for Calabi-Yau modular forms t_0, t_4 interpreted as holomorphic functions on $\mathbb{H}$, see (10.3). Recall that in the elliptic curve case the periods of holomorphic 1-forms over topological cycles never vanish, and a similar construction of the Eisenstein series E_4 and E_6 in terms of elliptic periods, see §2.6, shows that E_4 and E_6 can only have zeros in the elliptic points $(-1)^{\frac{1}{2}}$ and $(-1)^{\frac{1}{3}}$, respectively.

11.2 Hecke operators

Modular forms as generating functions count very unexpected objects beyond the scope of analytic number theory. The Shimura-Taniyama conjecture, now the modularity theorem, states that the generating function for counting $\mathbb{F}_p$-rational points of an elliptic curve over $\mathbb{Z}$ for different primes p, is essentially a modular form (see [Wil95] for the case of semi-stable elliptic curves and [BCDT01] for the case of all elliptic curves). Monstrous moonshine conjecture, now Borcherds theorem, relates the coefficients of the j-function with the representation dimensions of the monster group (see [Bor92] and the expository article [Gan06]). In both these amazing applications, Hecke operators play an essential role. Since the first drafts of the present text was written, the author was very eager to find Hecke operators for Calabi-Yau modular forms, however, all his attempts failed. In this and the next section, we are going to explain this.

Hecke operators T_n, $n \in \mathbb{N}$ acts on the space of full modular forms of a fixed weight and it leaves the space of cusp forms invariant. These are modular forms

$f=\sum_{n=0}^{\infty} f_n q^n$ with $f_0=0$. If a non-zero cusp form f with $f_1=1$ is a simultaneous eigenform of all Hecke operators then f_n's are multiplicative, that is, $f_{nm}=f_n f_m$ for n and m relatively prime numbers. This is compatible with the fact that for weight k cusp forms f for any Fuchsian group, we have the bound $f_n=O(n^{k/2})$, see for instance [Miy89, Zag08]. In the case of differential Calabi-Yau modular forms the coefficients of q-expansions are expected to grow exponentially which is not compatible with the above feature of modular forms. For instance, in [BCOV94] it has been argued that the coefficients of F_g grow like $a\cdot n^{2g-3}(\log n)^{2g-2}b^n$, where a,b are two constants depending on g and the underlying Calabi-Yau threefold.

Another way to predict the existence of Hecke operators is as follows. Any modular form $f=\sum_{n=0}^{\infty} f_n q^n$ for $\mathrm{SL}(2,\mathbb{Z})$ is in the polynomial ring $\mathbb{C}[E_4,E_6]$. Now, using Hecke operators one can say more. The function $f(q^n)$, $n\geq 1$ is an algebraic integer over the ring $\mathbb{C}[E_4,E_6]$, that is, there is a polynomial $P_n\in\mathbb{C}[X,Y,Z]$ which is monic in X and $P_n(f(q^n),E_4,E_6)=0$. Don Zagier (and many others) have written computer codes to detect such algebraic dependencies. During a 2012 visit to Max-Planck institute, the author discussed this issue with Don Zagier in the case of modular forms. The conclusion of our discussions and his computer search was that most probably such phenomena does not occur for differential Calabi-Yau modular forms.

A geometric recipe for producing Hecke operators uses isogenies of elliptic curves, however, this also seems to fail in the case of mirror quintics. This is explained in the next section.

11.3 Maximal Hodge structure

In this section we gather some problems with Hodge theoretic flavors. For missing definitions the reader is referred to [Lew99].

Problem 3 Let M be a smooth compact surface and let X be a mirror quintic. Any holomorphic map $i: M\to X$ induces the zero map in third homologies, that is, $i_*: H_3(M,\mathbb{Z})\to H_3(X,\mathbb{Z})$ is the zero map.

The equality $\int_{i_*(\delta)}\eta=\int_\delta i^*\eta$ implies that if Problem 3 is wrong then we have zeros for the periods of η. Problem 3 is equivalent to

$$H^3(X,\mathbb{Q})\cap(H^{21}\oplus H^{12})=\{0\},$$

where H^{ij}'s are pieces of the Hodge decomposition of $H^3(X,\mathbb{C})$. A weaker version of this statement is to say that the intersection $H^3(W,\mathbb{Q})\cap(H^{21}\oplus H^{12})$ has no maximal sub Hodge structure. If the generalized Hodge conjecture is true then this is equivalent to Problem 3. The variety $\mathsf{X}_{1,\infty}$ is obtained by the quotient of the Fermat variety and hence its periods can be computed in terms of the values of the Γ function on rational numbers, see Deligne's lecture notes in [DMOS82] or [Mov11b]. Therefore, it seems to me that analyzing these problems for $z=\infty$ will help us.

The author learned from Duco van Straten the following natural fibration of mirror quintics with Abelian surfaces. Recall that X_z is given by the equation (4.21). We introduce a new variable $y_0 = (y_1y_2y_3y_4)^{-1}$ and define $M_{z,s}$ in the following way:

$$M_{z,s}: \begin{cases} y_0 + y_1 + y_2 + y_3 + y_4 = 5z^{-\frac{1}{5}}, \\ y_0y_1y_2y_3y_4 = 1, \\ y_0y_1 + y_1y_2 + y_3y_4 + y_4y_0 = s. \end{cases} \tag{11.1}$$

The first two equations give us X_z and so we have the fibration of X_z's with $M_{z,s}$'s. The statement of Problem 3 is true for $M = M_{z,s}$. This follows from the fact that any linear combination $n_0\tau_0 + n_1 + n_1\tau_2 + n_3\tau_3$ with $n_i \in \mathbb{C}$ cannot be identically zero. To see this, take τ_0 an independent variable and other τ_i's as functions in τ_0 as in §2.10.

The Hecke operators act on the algebra of modular forms and this is the starting point of many number theoretic applications. After introducing modular and quasi-modular forms in the framework of Algebraic Geometry of elliptic curves, see for instance [Mov08a, Mov12b], the Hecke operators are constructed using an algebraic cycle in the product of two elliptic curves (which forces them to be isogenous). We have introduced differential Calabi-Yau modular forms for mirror quintics, and we may also wish to find Hecke operators in this case using the same geometric recipe. This goes as follows. Let $V_i(\mathbb{Q}) := H^3(\mathsf{X}_{1,z_i}, \mathbb{Q})$, $i = 1,2$ be two polarized Hodge structure arising from mirror quintics. In the rank 16 $\mathbb{Q}$-module $W(\mathbb{Q}) := V_1(\mathbb{Q}) \times V_2(\mathbb{Q})$ we consider the product Hodge decomposition. Its Hodge numbers are $1,2,3,4,3,2,1$.

Problem 4 *Is the following statement true? For V_1 fixed as above, there is always V_2 such that*

$$W(\mathbb{Q}) \cap (H^{51} \oplus H^{42} \oplus \cdots \oplus H^{24} \oplus H^{15}) \tag{11.2}$$

has a non-zero sub Hodge structure.

If the above statement is true we can consider the maximal sub Hodge structure in (11.2). By the generalized Hodge conjecture this comes from the fifth cohomology of a codimension one algebraic cycle $M \subset \mathsf{X}_{1,z_1} \times \mathsf{X}_{1,z_2}$. A simple dimension calculation may indicate that the statement in Problem 4 is not true. Let us assume that such a non-zero sub Hodge structure exists and is of dimension n. The wedge product $\wedge^n W(\mathbb{Q})$ carries a Hodge structure of weight $6n$ and it has a non-zero Hodge cycle in its $H^{3n,3n}$ piece. Such a Hodge cycle is defined by $a = \sum_{i<3n} \dim(H^{i,6n-i})$ equalities. If all these equalities are independent from each other, then such a Hodge cycle exists if and only if $a = 1$. That is, the Hodge numbers of the wedge product $\wedge^n W(\mathbb{Q})$ must be of the form $\ldots, 0, 1, x, 1, 0, \ldots$ which cannot happen.

11.4 Monodromy

In §2.7 we have fixed a basis δ of $H_3(\mathsf{X}_{1,z},\mathbb{Z})$ such that the monodromy representation for anticlockwise paths around the singularities $z=0$ and $z=1$ are respectively given by M_0 and M_1 in (2.26). The authors in [BT14] have proved that the monodromy group is free and so it is a subgroup of $\mathrm{Sp}(4,\mathbb{Z})$ of infinite index. Their method is based on the Ping-Pong lemma. It is not at all clear whether this lemma could be related to the geometry of the underlying Calabi-Yau threefolds $\mathsf{X}_{1,z}$. It would be interesting to prove this statement without the explicit computations of M_0 and M_1 and using the algebraic geometric tools for the family $\mathsf{X}_{1,z}$.

11.5 Torelli problem

S. Usui in [Usu08] has proved that the generic Torelli problem is true for mirror quintics. His proof uses the partial compactification of the moduli of log Hodge structures developed by him and Kato in [KU02]. However, note that the global Torelli for mirror quintic Calabi-Yau threefolds is still open.

Let D be the Griffiths period domain associated to Hodge structures on the third cohomology of mirror quintic Calabi-Yau threefolds, see §4.4. We have $\dim(D)=4$ and D has natural coordinates given by $\tau_i,\ i=0,1,2,3$ as independent variables.

Problem 5 *(Global Torelli problem) The period map*

$$\mathbb{P}^1-\{0,1\}\to \mathrm{Sp}(4,\mathbb{Z})\backslash D, \tag{11.3}$$

$$z\mapsto \textit{Hodge structure of } H^3(\mathsf{X}_{1,z},\mathbb{Z}),$$

is one to one.

The idea is to construct a partial compactification of $\mathrm{Sp}(4,\mathbb{Z})\backslash D$ and prove that the period map extends to $\mathbb{P}^1$. This together with the local Torelli gives a solution to the above problem. We try to develop this idea not using D but the image M of the period map.

Let M be the set of all Hodge structures arising from the third cohomology of mirror quintic Calabi-Yau threefolds, that is, M is the image of the period map (11.3). We add two other points to M:

$$\bar{M}:=M\cup\{\infty_0,\infty_1\}$$

and we would like to construct a holomorphic structure on $\bar{M}$ such that the extended period map

$$\mathsf{P}:\mathbb{P}^1\to\bar{M},\ z\mapsto \text{Hodge structure of } H^3(\mathsf{X}_{1,z},\mathbb{Z}),\ z\neq 0,1,$$

$$0\mapsto\infty_0,\ 1\mapsto\infty_1$$

is a biholomorphism. According to the local Torelli, the period map is one to one in small neighborhoods in $\mathbb{P}^1 - \{0,1\}$. Therefore, we can use P as a local chart outside 0 and 1. However, this will not help too much. We have to find charts for $\bar{M}$ using periods. The candidates for charts around ∞_i, $i = 0,1$ are as follows. There is a unique basis δ_i, $i = 1,2,3,4$ of $H_3(\mathsf{X}_{1,z},\mathbb{Z})$ which is characterized by the property that the monodromy around 0 and 1 is given by the matrices M_0 and M_1, respectively, see (2.26). We choose the following coordinate system around ∞_0 and ∞_1:

$$q_0(w) := e^{2\pi i \tau_0(w)}, \quad w \in M, \quad q_0(\infty_0) = 0,$$

$$q_1(w) := e^{2\pi i \theta_0(w)}, \quad w \in M, \quad q_1(\infty_1) = 0.$$

Here, w is a Hodge structure arising from a mirror quintic X such that in some symplectic basis of $H_3(X,\mathbb{Z})$ and a basis of $H^3_{\mathrm{dR}}(X)$, the corresponding period matrix is of the τ-format (4.5) and $\tau_0(w)$ is in a neighborhood of $i\infty$. We have also $\theta_0(w) = \frac{1}{\tau_2(w)}$. For $A = [a_{ij}] \in \mathrm{Sp}(4,\mathbb{Z})$ let

$$A(\tau_0) := \frac{a_{11}\tau_0 + a_{12} + a_{13}\tau_1 + a_{14}\tau_2}{a_{21}\tau_0 + a_{22} + a_{23}\tau_1 + a_{24}\tau_2}, \quad A(\theta_0) := \frac{a_{21}\theta_1 + a_{22}\theta_0 + a_{23}\theta_2 + a_{24}}{a_{41}\theta_1 + a_{42}\theta_0 + a_{43}\theta_2 + a_{44}}.$$

A solution to global Torelli problem requires the following:

Problem 6 *There is a real number $r_0 >> 0$ such that for all $A \in \mathrm{Sp}(4,\mathbb{Z})\backslash\{M_0^i,\ i \in \mathbb{Z}\}$ and $\tau_0 \in \mathbb{C}$ with $\mathrm{Im}(\tau_0) > r$ we have $\mathrm{Im}(A(\tau_0)) < r_0$. In a similar way, there is a real number $r_1 >> 0$ such that for all $A \in \mathrm{Sp}(4,\mathbb{Z})\backslash\{M_1^i,\ i \in \mathbb{Z}\}$ and $\theta_0 \in \mathbb{C}$ with $\mathrm{Im}(\theta_0) > r_1$ we have $\mathrm{Im}(A(\theta_0)) < r_1$.*

The above statement implies that the analytic continuation of the meromorphic map $\tau_0 : \mathbb{H} \to \mathbb{C}$ from a point with values near $i\infty$, remains near $i\infty$ if and only if we are doing monodromy around the MUM singularity $z = 0$. Note that the same problem in the case of elliptic curves follows from the well-known properties of the action of $\mathrm{SL}(2,\mathbb{Z})$ on the upper half plane.

11.6 Monstrous moonshine conjecture

The parameter $j_* = z^{-1} = \frac{t_0^5}{t_4}$ classifies the Calabi-Yau varieties of type (3.1), that is, each such Calabi-Yau variety is represented exactly by one value of j_* and two such Calabi-Yau varieties are isomorphic if and only if the corresponding j_* values are equal. This is similar to the case of elliptic curves which are classified by the classical j-function:

$$j = 1728\frac{E_4^3}{E_4^3 - E_6^2} = q^{-1} + 744 + 196884q + 21493760q^2 + 864299970q^3 + \cdots. \tag{11.4}$$

The beautiful story of monstrous moonshine conjecture starts with the key observation by John McKay in 1978 that 196884 is one more than the number of dimensions in which the monster group can be most simply represented. Later, John Conway and Simon Norton formulated the monstrous moonshine conjecture and Richard Borcherds gave a solution to it, see [Gan06] for an expository account of all these. What we would like to emphasize here is a gap of more than hundred years between the first computation of the coefficients of j and the observation of McKay. In a private conversation, McKay pointed out that after this observation he started to look for some old texts on j to see whether this was an accident or not. This urges us to think that there might be similar phenomena for j_*, however, our contemporary mathematics is not enough developed to see such connections, that is, the similar object to monster group in our case is not yet studied or discovered. Therefore, we have just computed the q-expansions of j_* and we hope that the future works in mathematics will reveal a meaning for this:

$$\begin{aligned} 3125 \cdot j_* = & \frac{1}{q} + 770 + 421375q + 274007500q^2 + 236982309375q^3 \\ & +251719793608904q^4 + 304471121626588125q^5 \\ & +401431674714748714500q^6 + 562487442070502650877500q^7 \\ & +824572505123979141773850000q^8 \\ & +1013472859153384775272872409691q^9 + \cdots . \end{aligned}$$

The coefficient 3125 is chosen in such a way that all the coefficients of $q^i, i \leq 9$ in $3125 \cdot j_*$ are integers and all together are relatively prime. Note that the moduli parameter j_* in our case has two cusps ∞ and 1, that is, for these values of j_* we have singular fibers. Our q-expansion is written around the cusp ∞.

11.7 Integrality of instanton numbers

The integrality of coefficients of differential Calabi-Yau modular forms attached to 14 cases of hypergeometric functions has been proved in [KRNT], see Appendix C for a discussion of this topic. The main ingredient of the proof is the so-called Dwork method which is available only for hypergeometric functions, and therefore, for other Calabi-Yau equations or varieties with non-hypergeometric solutions or periods as in [GAZ10], the integrality is still a conjecture.

Problem 7 *Prove a similar statement as in Theorem 5 for all the Calabi-Yau equations in [GAZ10].*

A solution to this problem would require generalizations of Dwork method beyond the hypergeometric functions. This in itself seems to be a hard job.

The integrality of coefficients of differential Calabi-Yau modular forms do not imply the integrality of the instanton numbers n_d in the Yukawa coupling (2.30). This is because the Yukawa coupling is written in the Lambert series format and then we have divided the d-th coefficient by d^3. In the announcement [KSVv3] the authors have claimed that they have solved this problem for mirror quintics, however, a complete proof has not been appeared and apparently their method is erroneous or incomplete. Therefore, the following is still open for mirror quintics.

Problem 8 *Prove that the instanton numbers n_d, calculated recursively either by the differential equation (2.9) or by the period manipulation in §2.15, are integers.*

11.8 Some product formulas

In this section we aim to find some conjectural statements for the Fourier expansions of differential Calabi-Yau modular forms in the case of mirror quintics. Our main strategy is to write the Lambert series of a Fourier expansion and to see whether the n-th coefficient is divisible by some power of n or not.

Recall the genus one topological partition function $\mathsf{F}_1^{\mathrm{alg}}$ in (6.2) and (6.3). Inside the logarithm used in $\mathsf{F}_1^{\mathrm{alg}}$ we have exactly the discriminant of the moduli space T. However, we do not have a control of the exponent of each factor. We would like to analyze the product formula for each piece in the left hand side of the equality (6.3). We write the logarithmic derivative of each factor $X = t_4, t_4 - t_0^5, t_5$ in the form

$$\dot{\ln} X = a_0 - \sum_{n=1}^{\infty} a_n \frac{nq^n}{1-q^n}, \quad a_i \in \mathbb{Q}.$$

The main reason for writing in this format is that we get the product formula for X:

$$X = a \cdot q^{a_0} \prod_{n=1}^{\infty} (1-q^n)^{a_n}, \quad a \in \mathbb{Q}.$$

For $X = t_4^{\frac{1}{10}}$ we get

$$a_0 = 1/10, a_n = 17, 5584, 4087136, 4041677110, \ldots$$

[Supp Item 41], for $X = (t_4 - t_0^5)^{\frac{1}{5}}$ we get

$$a_0 = 0, a_n = 505, 481260, 625630915, 919913221900, \ldots$$

[Supp Item 42] and for $X = t_5^{\frac{1}{25}}$ we get

$$a_0 = 0, a_n = 75, 74965, 99623375, 148496906530, \ldots$$

[Supp Item 43]. Computing a_n for big n we get the following:

Problem 9 *All the a_n corresponding to $t_4^{\frac{1}{10}}, (t_4 - t_0^5)^{\frac{1}{5}}, t_5^{\frac{1}{25}}$ are integers.*

Comparing the coefficients a_1's in $\mathsf{F}_1^{\mathrm{hol}}, t_5, t_4, t_4-t_0^5$ we get the identity: $25\cdot 17\cdot 10-5\cdot 5\cdot 505+6\cdot 25\cdot 75=2875$. Knowing that 2875 is the number of rational curves in a generic quintic, one might look for an enumerative meaning of this. We may try to write the Lambert series format for other modular type functions. Among $(5t_0)^k = 1+\cdots$ we find that only for $k=2,3$ we get a nice Lambert series:

$$(5t_0)^k = 1+\prod_{n=1}^{\infty} a_n \frac{q^n}{1-q^n}.$$

For $k=1$ apparently infinitely many primes appear in the sequence a_n/n. For $k=2$ we observe that $\frac{a_n}{240n}, n\in\mathbb{N}$ is given by

$$1, 117, 46208, 31934160, 28568471125, \ldots$$

[Supp Item 44] and for $k=3$ we observe that $\frac{a_n}{60n^2}$ is given by

$$1, 6, 441, 109616, 54828405, 38606737350, \ldots$$

[Supp Item 45] which are apparently all integers. For $(5t_0)^4$ the coefficient a_n is conjecturally divisable by n and not n^2. For $(5t_0)^5$ the coefficient a_n is not divisible by n. All these affirmations are checked up to 20 coefficients. One may also think that the propagators might have nice Lambert series structures. For instance, take

$$S^{zz} = \frac{-1}{a_5}\frac{u_6}{u_5} = -(z-1)\frac{u_6}{u_5} = -\frac{t_4-t_0^5}{t_0^5}\frac{5^5(t_4-t_0^5)}{t_5 t_0^3}\left(\frac{t_6 t_0^4}{5^5(t_4-t_0^5)} - 3t_5\right) = -\frac{(t_4-t_0^5)t_6}{t_0^4 t_5},$$

where the last equality is written up to an ambiguity (expressions in t_0 and t_4). We do not get any divisibility property for S^{zz}. The following question arises in a natural way. Is it possible to fix the ambiguity of propagators by requiring that the n-coefficient in the Lambert series format is divisible by n? For the computer codes used in the computations of this section see the author's webpage [Supp Item 46].

A product formula for the quantity $Z := \exp\sum_{g=1}^{\infty} \mathsf{F}_g^{\mathrm{alg}} \lambda^{2g-2}$ has been obtained in [GVv1] using the so-called Gopakumar-Vafa invariants. Such invariants are conjectured to be integers. See also [KKRS05], page 42 for further discussion of this topic.

11.9 A new mirror map

The computation of this section has been suggested/motivated by Emanuel Scheidegger. Recall the following variable from (2.34):

$$\tau_3 = 5\tau_0 + \frac{5}{2} + \frac{1}{(2\pi i)}\sum_{n=1}^{\infty}\Big(\sum_{d|n} n_d d^3\Big)\frac{e^{2\pi i \tau_0 n}}{n}.$$

We immediately realize that we can use $q = e^{2\pi i \tau_0}$ as well as $q_3 := q = ^{\frac{2\pi i \tau_3}{5}}$ as a coordinate system around $z = 0$. We have $q_3 = -q - 575q^2 + \cdots$. From this and the first coefficients of t_i's in q coordinate we get the first coefficients of t_i's in the new coordinate q_3:

$$t_0 = \frac{1}{5} - 24q + \cdots,\ t_1 = -25 + 2250q + \cdots, \tag{11.5}$$

$$t_2 = -35 + 5350q + \cdots,\ t_3 = -6 + 355q + \cdots,$$

$$t_4 = -q + \cdots,\ t_5 = -1 - 1875q + \cdots,$$

$$t_6 = -15 - 4675q + \cdots.$$

We know that

$$q_3 \frac{\partial}{\partial q_3} = \frac{1}{\mathsf{Y}} 5q \frac{\partial}{\partial q}, \quad \text{where } \mathsf{Y} = \frac{-5^{11}(t_4 - t_0^5)^2}{t_5^3}$$

and so the differential equation of t_i's with the above derivation becomes $\frac{1}{\mathsf{Y}}\mathsf{R}_0$. Therefore, we have

$$\mathsf{R}: \begin{cases} \dot{t}_0 = T(6\cdot 5^4 t_0^5 + t_0 t_3 - 5^4 t_4), \\ \dot{t}_1 = T(-5^8 t_0^6 + 5^5 t_0^4 t_1 + 5^8 t_0 t_4 + t_1 t_3), \\ \dot{t}_2 = T(-3\cdot 5^9 t_0^7 - 5^4 t_0^5 t_1 + 2\cdot 5^5 t_0^4 t_2 + 3\cdot 5^9 t_0^2 t_4 + 5^4 t_1 t_4 + 2t_2 t_3), \\ \dot{t}_3 = T(-5^{10} t_0^8 - 5^4 t_0^5 t_2 + 3\cdot 5^5 t_0^4 t_3 + 5^{10} t_0^3 t_4 + 5^4 t_2 t_4 + 3t_3^2), \\ \dot{t}_4 = T(5^6 t_0^4 t_4 + 5 t_3 t_4), \\ \dot{t}_5 = T(-5^4 t_0^5 t_6 + 3\cdot 5^5 t_0^4 t_5 + 2t_3 t_5 + 5^4 t_4 t_6), \\ \dot{t}_6 = T(3\cdot 5^5 t_0^4 t_6 - 5^5 t_0^3 t_5 - 2t_2 t_5 + 3t_3 t_6) \end{cases} \tag{11.6}$$

with

$$\dot{*} := q_3 \frac{\partial *}{\partial q_3}, \quad T := \frac{t_5^2}{-5^{11}(t_4 - t_0^5)^2}.$$

As in the case of R_0, this differential equation together with the initial values (11.9) gives us the Fourier expansion of t_i's in q_3. We compute the first 20 coefficients and get the conclusion:

Problem 10 *All the coefficients of q_3^i, $i \le 1$ in the series*

$$\frac{1}{24}t_0,\ \frac{1}{750}t_1,\ \frac{1}{50}t_2,\ \frac{1}{5}t_3,\ t_4,\ \frac{1}{125}t_5,\ \frac{1}{25}t_6$$

[Supp Item 47] are integers.

If we write all t_i's in the Lambert series format $t_i = t_{i,0} + \sum_{n=1}^{\infty} a_{i,n} \frac{q_3^n}{1-q_3^n}$ we realize that for none of t_i's we can factor out n, that is, $a_{i,n}/n$ in general is a rational number. Emanuel Scheidegger observed the following structure:

$$5\frac{t_5^3}{-5^{11}(t_4-t_0^5)^2}=1+25\sum_{n=1}^{\infty}a_nn^2\frac{q_3^n}{1-q_3^n} \tag{11.7}$$

where $a_n, n=1,2,3,\dots$ are given by

$$23,-3147,198328,145843715,-91484169495,30431124288600,\dots$$

[Supp Item 48].

Problem 11 *All a_n's in* (11.7) *are integers.*

For the computer codes of the present section see the author's webpage [Supp Item 49].

11.10 Yet another coordinate

Comparing the Ramanujan differential equation with R_0 in (2.9), one immediately realize a difference, namely the former is polynomial, however, the later is meromorphic in t_5. The locus $t_5=0$ in T is also part of the discriminant locus, however, it does not correspond to degeneration of mirror quintics but the degeneration of the elements of the de Rham cohomologies. Don Zagier tried to rewrite new generators such that the differential equation (2.9) becomes polynomial, see (2.12). In this section we show that this is partially possible when we change the derivation.

Let us take a coordinate system Q around $z=0$ such that

$$t_5(q)5q\frac{\partial}{\partial q}=a\cdot Q\frac{\partial}{\partial Q}$$

for some constant a. Since $t_5=-1+O(q)$ we find out that up to multiplication by a constant $Q=q^{\frac{-a}{5}}(1+\cdots)$. Since we assume that Q is a coordinate system, we have $a=-5$. Finally we get the following ordinary differential equation between $t_i,\ i=0,1,2,3,4$:

$$\mathsf{R}:\ \begin{cases} \dot t_0=6\cdot 5^4t_0^5+t_0t_3-5^4t_4,\\ \dot t_1=-5^8t_0^6+5^5t_0^4t_1+5^8t_0t_4+t_1t_3,\\ \dot t_2=-3\cdot 5^9t_0^7-5^4t_0^5t_1+2\cdot 5^5t_0^4t_2+3\cdot 5^9t_0^2t_4+5^4t_1t_4+2t_2t_3,\\ \dot t_3=-5^{10}t_0^8-5^4t_0^5t_2+3\cdot 5^5t_0^4t_3+5^{10}t_0^3t_4+5^4t_2t_4+3t_3^2,\\ \dot t_4=5^6t_0^4t_4+5t_3t_4 \end{cases} \tag{11.8}$$

with $\dot{} = -5\cdot Q\frac{\partial}{\partial Q}$ and

$$t_0=\frac{1}{5}-24q+\cdots,\ t_1=-25+2250q+\cdots,\ t_2=-35+5350q+\cdots, \tag{11.9}$$

$$t_3=-6+355q+\cdots,\ t_4=-q+\cdots$$

[Supp Item 50]. Unfortunately we do not get q-expansion with integer coefficients. The primes in the denominators grow slowly. For the computer codes of the present section see the author's webpage [Supp Item 51].

11.11 Gap condition

We discuss the gap condition for topological string partition functions in the framework of the moduli T and its Gauss-Manin connection. Recall the notation y_{ij} from §4.15. Let us take a mirror map $q := \frac{y_{11}}{y_{21}}$ with

$$\langle y_{11}, y_{21} \rangle = 0, \quad q = O((z-1)). \tag{11.10}$$

The gap condition for the topological string partition function says that

$$\mathsf{F}_g^{\mathrm{hol}} = \frac{(-1)^g B_{2g}}{2g(2g-2)} \frac{1}{q^{2g-2}} + O(q^0),$$

see for instance [HKQ09]. In this section we are mainly interested on the vanishing of the coefficients of q^i, $i = -1, -2, \ldots, 2g-3$ in the q-expansion of $\mathsf{F}_g^{\mathrm{hol}}$.

Problem 12 *The gap condition is independent of the mirror map with the conditions* (11.10).

Note that for the choice of mirror map in §5.4 we have four free parameters. The above problem has been solved partially in [HKQ09], however, their argument is incomplete. Below, we give a computational proof to the above problem in the genus two and three case.

We analyze the gap condition for $\mathsf{F}_2^{\mathrm{hol}}$. Recall from §5.4 that we can use the vector field R_0 and write down the q-expansion of all t_i's around the conifold. We are going to use the change of variables used in §5.5. Since there are two canonical $\mathbb{C}^*$-action on T, we can set $t_{0,0} = 1$, $t_{5,1} = e_3 = 1$. In this way the derivation is $5^5 \frac{\partial}{\partial q}$. All the series t_i has coefficients in $\mathbb{Q}[e_0, e_2, e_3]$, where we have used the notation $e_0 = t_{1,0}$. We write $\mathsf{F}_2^{\mathrm{hol}}$ in the new variables in §5.5 [Supp Item 52]. The ambiguity of $\mathsf{F}_2^{\mathrm{hol}}$ comes from the monomials

$$t_0^{12}/t_4^2, \;\; t_4 t_0^7/t_4^2, \;\; t_4^2 t_0^2/t_4^2.$$

The last monomial is holomorphic at $q = 0$. Therefore, only the first two monomials can contribute to the gap condition. Let us write

$$\mathsf{F}_2^{\mathrm{hol}} := F_2 + X t_0^{12}/t_4^2 + Y t_4 t_0^7/t_4^2,$$

where F_2 is the part of $\mathsf{F}_2^{\mathrm{hol}}$ free of t_0^{12}/t_4^2 and $t_4 t_0^7/t_4^2$. We plug the formal power series t_i in $\mathsf{F}_2^{\mathrm{hol}}$ and find out that

$$\mathsf{F}_2^{\text{hol}} := \frac{X}{25e_0^2e_1^2 - 50e_0e_1e_2 + 25e_2^2}\frac{1}{q^2} + \frac{2016X - 1440Y + 1015625}{7200e_0e_1 - 7200e_2}\frac{1}{q} + O(q^0).$$

The coefficient of $\frac{1}{q}$ is zero and so $2016X - 1440Y + 1015625 = 0$ which is independent of the parameters e_0, e_1, e_2. For the computer code used in this computation see the author's webpage [Supp Item 53].

In a similar way, we have verified the following gap condition for $\mathsf{F}_3^{\text{hol}}$:

$$\mathsf{F}_3^{\text{hol}} = \frac{(6103515625/1008)}{625e_0^4e_1^4 - 2500e_0^3e_1^3e_2 + 3750e_0^2e_1^2e_2^2 - 2500e_0e_1e_2^3 + 625e_2^4}\frac{1}{q^4} + *\frac{1}{q} + O(q)$$

for arbitrary mirror map as in the previous section [Supp Item 54]. Due to the limit of my computer I was not able to verify the gap condition for $\frac{1}{q}$. For particular choices of e_0, e_1, e_2 one can verify the full gap condition.

11.12 Algebraic gap condition

In this section we use the new variables introduced in §5.5. Motivated by the gap condition we define

Definition 6 *A formal power series* $\sum_{k=-n}^{\infty} a_k q^k$, $a_{-n} \neq 0$ *satisfies the (analytic) gap condition if* $a_{-n+1} = \cdots = a_{-1} = 0$.

The above definition in our context highly depends on the choice of the mirror map q. We would like to translate this definition to the algebraic context of the moduli space T and the vector field R_0. First note that if a formal power series f with pole order n at $q = 0$ satisfies the gap condition then

$$f = \frac{\partial^{n-1} g}{\partial q^{n-1}}$$

for some formal power series g with pole order one at $q = 0$. The derivation $\frac{\partial}{\partial q}$ is translated into the vector field R_0 in T and so it is natural to define.

Definition 7 A rational function $F \in \mathbb{C}[t, \frac{1}{t_5 t_4}]$ has an order at least n along the conifold singularity if it is a $\mathbb{C}$-linear combination of monomials t^i with $i_2 + i_3 + i_4 + i_5 + i_6 \geq n$. It has the order n if, further, it is not of order $n+1$. It has a pole order n if it has order $-n$.

Definition 8 Let $p \in Sin_{1,c}$, $p \neq 0$. A rational function $F \in \mathbb{C}[t, \frac{1}{t_5 t_4}]$ with pole order n along the conifold satisfies the algebraic gap condition if

$$F = \mathsf{R}_0^{n-1} P(t)$$

for some formal power series P around p and with possible poles along along $t_4 = t_5 = 0$.

Proposition 29 *The algebraic gap condition implies the analytic gap condition.*

Proof. The vector field R_0 acts on the formal power series as $\frac{\partial}{\partial q}$.

Integrating along vector fields: Let V be a holomorphic vector field in $(\mathbb{C}^n, 0)$ with a singularity at 0 and with a non-zero linear part. Let also f be a meromorphic function in $(\mathbb{C}^n, 0)$ with possible poles along $z_i = 0$'s. We want to find another meromorphic function g such that

$$Vg = f \tag{11.11}$$

that is we want to integrate f along V. The linear part of V plays an important role on the integrability of the function f. Let us write

$$V = V_1 + V_2 + \cdots, \; f = f_n + f_{n+1} + \cdots$$

and so $g = g_n + g_{n+1} + \cdots$. Writing the homogeneous pieces of (11.11) we get

$$\begin{aligned} V_1 g_n &= f_n, \\ V_1 g_{n+1} + V_2 g_n &= f_{n+1}, \\ &\vdots \\ V_1 g_{n+m} + V_2 g_{n+m-1} + \cdots + V_{m+1} g_n &= f_{n+m}. \end{aligned}$$

This means that if we are able to integrate along the linear part V_1 of V then we can integrate f along V formally. We have

Proposition 30 *If the liner part of V is of the form*

$$V_1 = x_1 \frac{\partial}{\partial x_1} + x_2 \frac{\partial}{\partial x_2} + \cdots + x_n \frac{\partial}{\partial x_n},$$

then f is integrable along V if either $n \geq 0$, or $n < 0$ and in the $-n$ step of the above process

$$V_1 g_0 + V_2 g_{-1} + \cdots + V_{-n+1} g_n = f_0 \tag{11.12}$$

holds automatically.

Proof. Note that for a homogeneous rational function g of degree d we have $V_1 g = dg$ and so for $d = 0$ we have $V_1 g = 0$. This means that $V_1 g_0 = 0$ and we cannot compute g_0 through the equality (11.12).

Analyzing the algebraic gap condition: We would like to examine the algebraic gap condition around

$$p = (p_0, p_1, 0, 0, 0, 0, 0) \in Sin_{1,c}.$$

We consider the usual weights $weight(t_i) = 1$, $i = 0, 1, 2, \ldots, 6$ and set $V := \mathsf{R}_0$. Let us calculate the linear part V_1 of V around p. It is

$$\frac{1}{t_5}[\frac{\partial}{\partial t_0},\dots,\frac{\partial}{\partial t_6}]B[t_0-p_0,t_1-p_1,\dots,t_6-p_6]^{\mathrm{tr}},$$

where

$$B:=\begin{pmatrix} 0&0&0&p_0&-625&0&0\\ 0&0&0&p_1&390625p_0&0&0\\ 0&0&0&3125p_0^3&625p_1&0&0\\ 0&0&0&3125p_0^4&0&0&0\\ 0&0&0&0&3125p_0^4&0&0\\ 0&0&0&0&0&3125p_0^4&0\\ 0&0&0&0&0&3125p_0^3&0 \end{pmatrix}.$$

We find a matrix A such that ABA^{-1} is in the Jordan form:

$$A=\begin{pmatrix} 0&0&0&0&0&-\frac{1}{p_0}&1\\ 0&0&1&-\frac{1}{p_0}&-\frac{p_1}{5p_0^4}&0&0\\ 0&1&0&-\frac{p_1}{3125p_0^4}&-\frac{125}{p_0^3}&0&0\\ 1&0&0&-\frac{1}{3125p_0^3}&\frac{1}{5p_0^4}&0&0\\ 0&0&0&0&0&\frac{1}{p_0}&0\\ 0&0&0&0&\frac{1}{15625p_0^7}&0&0\\ 0&0&0&\frac{1}{3125p_0^4}&\frac{p_1}{15625p_0^7}&0&0 \end{pmatrix},$$

$$ABA^{-1}=\begin{pmatrix} 0&0&0&0&0&0&0\\ 0&0&0&0&0&0&0\\ 0&0&0&0&0&0&0\\ 0&0&0&0&0&0&0\\ 0&0&0&0&3125p_0^4&0&0\\ 0&0&0&0&0&3125p_0^4&0\\ 0&0&0&0&0&0&3125p_0^4 \end{pmatrix}.$$

The push-forward of the vector field V by

$$f:\mathbb{C}^7\to\mathbb{C}^7,\ f(t)=(x_0,x_1,\dots,x_6):=A(t-p)$$

is given by

$$\frac{1}{f_*t_5}[\frac{\partial}{\partial x_0},\dots,\frac{\partial}{\partial x_6}]ABA^{-1}[x_0,x_1,\dots,x_6]^{\mathrm{tr}}=\frac{3125p_0^4}{f_*t_5}W,$$

where

$$W:=x_4\frac{\partial}{\partial x_4}+x_5\frac{\partial}{\partial x_5}+x_6\frac{\partial}{\partial x_6}.$$

Note that the push-forward of the functions t_4 and t_5 are respectively

$$f_* t_4 = 15625 p_0^7 x_5,$$
$$f_* t_5 = p_0 x_4.$$

For the computer codes used in the present section see the author's webpage [Supp Item 55].

Proposition 31 *If the algebraic gap condition is valid for at least one point* $p = (p_0, p_1, 0, \ldots, 0) \in Sin_{1,c}$ *then it is valid for all* $p \in Sin_{1,c}$.

Proof. We have

$$x_4 = \frac{1}{p_0} t_5, \ x_5 = \frac{1}{15625 p_0^7} t_4, \ x_6 = \frac{1}{3125 p_0^4} t_3 + \frac{p_1}{15625 p_0^7} t_4.$$

Therefore, the linear part of R_0 at p does not contain $\frac{\partial}{\partial t_i}, \ i = 0, 1$ and during the integration process we need to consider only monomials in t_3, t_4, t_5 and so we need to consider formal power series in t_3, t_4, t_5 with coefficients which are polynomial in other t_i's. Note that $p_3 = p_4 = p_5 = 0$.

11.13 Arithmetic modularity

We are still far from a point in which our modular forms are considered as useful as classical modular forms. We called them differential Calabi-Yau modular forms or Calabi-Yau modular forms, however, the future development of the subject might choose a proper name for such objects. The author strongly believes that the appearance of fruitful applications of this kind of new modular forms is just a matter of time and more effort from both mathematicians and theoretical physicists. As we have seen, there are many open problems of complex analysis and topological nature, and as far as these problems are not solved, it seems hopeless to push forward more arithmetic oriented problems. However, we would like to point out some arithmetic problems which might be classified as a dream rather than a concrete problem.

The most celebrated application of modular forms is the so-called (arithmetic) modularity of elliptic curves. This was originally named as Shimura-Taniyama conjecture, and it states that the generating function for counting $\mathbb{F}_p$-rational points of an elliptic curve over $\mathbb{Z}$ for different primes p, is essentially a modular form (see [Wil95] for the case of semi-stable elliptic curves and [BCDT01] for the case of all elliptic curves). The generalization of this to other type of varieties is limited to K3 surfaces of high Picard rank and rigid Calabi-Yau varieties, see for instance Yui's expository article [Yui13] and the references therein. Even for Abelian varieties, for which we have a huge literature on their arithmetic aspects, we have partial results regarding the modularity theorem, see for instance [Rib04]. There is no example of a modular mirror quintic and this may indicate the fact that classical modular forms, and in general automorphic forms, are responsible for the modularity of a limited number of algebraic varieties, and beyond this, one needs new type of modular forms. Even though we do not contribute in this direction, this has been one of our main motivations in writing the present text. Counting points of Calabi-Yau threefolds over finite fields does not indicate any hint to this problem, see for instance [CdlORV03]. One might return back to the origins of the modularity theorem and search for some hints there. André Weil in the last paragraph of his article [Wei67], mentions that according to Hecke correspondence introduced in [Hec36], the L-function of an elliptic curve over $\mathbb{Q}$ must be a Mellin transform of a cusp form of weight 2, and he refers this to Shimura, see [Lan95b] for a historical account on this. Since the building blocks in the product formula of an L-function of a variety is derived from the zeta function of the same variety, one might hope that the L-function of a mirror quintic is a kind of Mellin transform of a differential Calabi-Yau modular form. However, in §2.10 we have seen that differential Calabi-Yau modular forms have strange functional equations with respect to the monodromy group Γ. In particular, no element of Γ acts on τ_0 as $\frac{-1}{\tau_0}$, a fact which is crucial in the analytic continuation of the Mellin transform of a cusp form. A more geometric way to say that an elliptic curve is modular is the existence of a morphism $X_0(N) \to E$, where N is the conductor of E and $X_0(N)$ is the modular curve parameterizing degree N isogenous pairs of elliptic curves. The generalization of this seems to be also difficult because there is no good generalization of isogeny in the case of Calabi-Yau threefolds, see §11.3. Before modularity theorem for elliptic curves was proved, we had algorithms for producing modular elliptic curves based on the period of modular forms, see [Cre97]. We do not see any modification of this approach adapted to the case of mirror quintics.

Appendix A
Second-order linear differential equations

In order to motivate the reader how we derive differential Calabi-Yau modular forms from linear differential equations, see §7, §8 and §9, we are going to discuss the case of classical quasi-modular forms and second-order differential equations. The presentation in §2 is suitable for cases in which we have an explicit family of elliptic curves, however, it can happen that we do not know the geometry and instead we know the corresponding Picard-Fuchs equations. In this chapter we discuss this. Since our main objective is to study differential Calabi-Yau modular forms we skip some details.

A.1 Holomorphic and non-holomorphic quasi-modular forms

Let us consider a family of elliptic curves E_z, $z \in S$, where S is a compact Riemann surface. We consider a meromorphic function z on S and use it as a coordinate system on S minus its poles and double zeros. We also consider a holomorphic family of regular 1-form $\eta = \eta_z,\ z \in S$ on E_z. If S is isomorphic to $\mathbb{P}^1$ we take z the meromorphic function obtained by this isomorphism. In this way η satisfies a second-order differential equation

$$\theta^2 - a_1(z)\theta - a_0(z) = 0, \tag{A.1}$$

where $\theta := z\frac{\partial}{\partial z}$. Let us define the 2×2 matrices

$$\mathrm{F} := [\langle \theta^{i-1}\eta,\ \theta^{j-1}\eta \rangle],\quad \mathrm{G} := [\langle \theta^{i-1}\eta,\ \overline{\theta^{j-1}\eta} \rangle], \tag{A.2}$$

$$\mathrm{X} := [x_{ij}],\quad x_{ij} := \theta^{j-1}x_{i1},$$

where x_{i1}, $i = 1,2$ are two linearly independent solutions of (A.1). A simple calculation shows that up to multiplication with a constant, which depends only on the triple $(E/S, \eta, z)$, we have

$$F_{12} := \langle \eta, \theta\eta \rangle = e^{\int a_1(z)\frac{dz}{z}}$$

and

$$F = \begin{pmatrix} 0 & F_{12} \\ -F_{12} & 0 \end{pmatrix}.$$

We have the equality $G^{tr}F^{-1}G = -\overline{F}$ which is equivalent to

$$G_{11}G_{22} - G_{12}G_{21} = |F_{12}|^2.$$

and so $\theta\bar{\theta}\log G_{11} = (\frac{|F_{12}|}{G_{11}})^2$. Any rational function in z, G_{11}, G_{21} (resp. z, x_{11}, x_{12}) for all choice of the meromorphic function z on S is called a non-holomorphic (resp. holomorphic) quasi-modular form attached to E/S. The holomorphic limit in this context is just the correspondence

$$z \leftrightarrow z, \quad G_{11} \leftrightarrow x_{11}, \quad G_{21} \leftrightarrow x_{12}.$$

For q-expansion of holomorphic modular forms one usually take a singularity $z = 0$ of (A.1) with one holomorphic solution $x_{11} = O(1)$ and one logarithmic solution $x_{21} = x_{11}\ln(z) + O(z)$. Then one makes the change of coordinates $z \mapsto q := e^{\frac{x_{21}}{x_{11}}}$.

Let α be a non-holomorphic quasi-modular form and consider the following differential equation:

$$\frac{\bar{\theta}X}{\overline{F_{12}}} = \alpha \tag{A.3}$$

for some unknown quantity X. We claim that (A.3) is equivalent to

$$G_{11}\frac{\partial X}{\partial G_{11}} + G_{21}\frac{\partial X}{\partial G_{21}} = 0, \quad \frac{F_{12}}{G_{11}}\frac{\partial X}{\partial G_{21}} = \alpha. \tag{A.4}$$

The basic idea behind the proof is as follows. We have

$$\begin{aligned}\bar{\theta}X &= G_{12}\frac{\partial X}{\partial G_{11}} + G_{22}\frac{\partial X}{\partial G_{21}} \\ &= \frac{G_{12}}{G_{11}}(G_{11}\frac{\partial X}{\partial G_{11}} + G_{21}\frac{\partial X}{\partial G_{21}}) + \overline{F_{12}}(\frac{F_{12}}{G_{11}})\frac{\partial X}{\partial G_{21}}.\end{aligned}$$

Since $\bar{z}$ and G_{12} are algebraically independent over the field of non-holomorphic quasi-modular forms, we conclude the proof. This is the heart of an argument used by Yamaguchi-Yau in [YY04] and Alim-Länge in [AL07] in the case of Picard-Fuchs equations coming from Calabi-Yau threefolds and in order to break the anti-holomorphic derivation into holomorphic ones.

A.2 Full quasi-modular forms

In order to get the Eisenstein series E_2, E_4, E_6 from the recipe of the previous section, we have to take a family of elliptic curves with $\mathrm{SL}(2,\mathbb{Z})$ monodromy. Therefore, we take the universal family of elliptic curves parametrized by their j-invariant:

$$E_j : f = 0, \tag{A.5}$$

$$f = y^2 + xy - x^3 + \frac{36}{j-1728}x + \frac{1}{j-1728},\ j \neq 0, 1728.$$

Note that in this family the fibers over $j = 0$ and $j = 1728$ are singular. One could take other families in which

$$E_0 : y^2 = x^3 + 1,\ E_{1728} : y^2 = x^3 - x$$

are smooth fibers. These elliptic curves have an automorphisms of order 3 and 2 respectively. The corresponding monodromies around these elliptic curves have orders 6 and 4 respectively. We consider $\mathbb{P}^1 = \mathbb{C} \cup \{\infty\}$ as the compactification of the moduli of elliptic curves with a cusp ∞. In the basis $[\alpha, \omega]^{\mathrm{tr}} = [\frac{dx\wedge dy}{df}, \frac{dx\wedge dy}{df}]^{\mathrm{tr}}$ of the de Rham cohomology $H^1_{\mathrm{dR}}(E_j)$, the Gauss-Manin connection is given by

$$\nabla_{\frac{\partial}{\partial j}} \begin{pmatrix} \alpha \\ \omega \end{pmatrix} = A \cdot \begin{pmatrix} \alpha \\ \omega \end{pmatrix}$$

where

$$A = \frac{1}{j(j-1728)} \begin{pmatrix} -432 & -60 \\ -(j-1728) & 432 \end{pmatrix}.$$

We use the parameter $z = \frac{1}{j}$ and write the Picard-Fuchs equation of α with respect to this parameter. The holomorphic differential 1-form α satisfies

$$-\frac{82944z - 372}{2985984z^3 - 3456z^2 + z} + \frac{1}{z}\partial_z + \partial_z^2 = 0.$$

This in θ-format is

$$\theta^2 - \frac{82944z^2 - 372z}{2985984z^2 - 3456z + 1} = 0.$$

For this example F_{12} is a constant. For the computer codes used in this computation see the author's webpage [Supp Item 56]. In order to find expressions of Eisenstein series in terms of z, G_{11} and G_{21}, we may use the following information. The discriminant $\Delta = E_4^3 - E_6^2$ vanishes of order one at $z = 0$ and it has no zeros elsewhere. The Eisenstein series E_4 and E_6 have zeros of order $\frac{1}{6}$ and $\frac{1}{4}$ at $z = \infty$ and $z = \frac{1}{1728}$, respectively. The Eisenstein series E_2 is the logarithmic derivative of Δ. The relations (A.4) are translated into

$$(2E_2\frac{\partial}{\partial E_2}+4E_4\frac{\partial}{\partial E_4}+6E_6\frac{\partial}{\partial E_6})(X)=0,$$

$$\frac{\partial}{\partial E_2}(X)=\alpha.$$

The final objective of this discussion would be to write down the anomaly equation of F_g's in the case of elliptic curves. This is missing in our arguments in §6.8.

Appendix B
Metric

In this chapter we describe a metric in the solutions of the Ramanujan vector field and its generalization (2.9) to the case of mirror quintics. First, we recall the usual presentation of this topic in the literature and then we adapt it to our case.

Let $\mathscr{M}$ be a moduli space of Calabi-Yau n-folds. By Bogomolov-Tian-Todorov's theorem, $\mathscr{M}$ is a smooth manifold. We take a local patch $(\mathbb{C}^m,0)$ of $\mathscr{M}$ and consider a family $X_z,\ z \in (\mathbb{C}^m,0)$ of Calabi-Yau n-folds. We denote by $(z_1,z_2,\dots,z_m)$ the coordinates system in $(\mathbb{C}^m,0)$. Let $\omega = \omega_z$ be a holomorphic n-form on X_z. We assume that ω depends holomorphically on z. The following Hermitian form on $\mathscr{M}$ has been used in the Physics literature

$$g := \sum_{i,j} \partial_{z_i}\bar{\partial}_{z_j} \log K \ \ dz_i \otimes d\bar{z}_j$$

where $K := \langle \omega, \bar{\omega} \rangle$ is the Kähler potential and $\langle \omega_i, \omega_j \rangle := \int_{X_z} \omega_i \wedge \omega_j$ for $\omega_i, \omega_j \in H^n_{\mathrm{dR}}(X_z)$, see for instance [CdlOGP91, BCOV93, BCOV94]. See also [LS04] and the references therein for some mathematically oriented discussions. It induces a metric on $\mathscr{M}$ which we call it the Kähler metric. This metric does not depend on the choice of ω and the coordinate system z. For instance, if we multiply ω with a holomorphic function $f(z)$ then

$$\partial_i \bar{\partial}_j \log(\langle f\omega, \overline{f\omega} \rangle) = \partial_i \bar{\partial}_j \log(\langle \omega, \overline{\omega} \rangle).$$

For $n \geq 2$ it is sometimes convenient to write down g in the following way. We define

$$\omega_i := \partial_{z_i}\omega - \frac{\langle \partial_{z_i}\omega, \bar{\omega} \rangle}{\langle \omega, \bar{\omega} \rangle}\omega, \tag{B.1}$$

$$f_i := \frac{\langle \partial_{z_i}\omega, \bar{\omega} \rangle}{\langle \omega, \bar{\omega} \rangle}. \tag{B.2}$$

Since $\langle \omega_i, \bar{\omega} \rangle = 0$, we conclude that ω_i in the $H^{n-1,1}$ piece of the Hodge decomposition of $H^n_{\mathrm{dR}}(X_z)$. We have

$$g = \sum_{i,j} \left(\frac{\langle \omega_i, \bar{\omega}_j \rangle}{\langle \omega, \bar{\omega} \rangle} + 2 f_i \bar{f}_j \right) dz_i \otimes d\bar{z}_j. \tag{B.3}$$

Note that the positivity condition on the polarized Hodge structure on $H^n_{\mathrm{dR}}(X_z)$ implies that $(-1)^{\frac{n}{2}} \langle \omega, \bar{\omega} \rangle > 0$ and the matrix $(-1)^{\frac{n-2}{2}} [\langle \bar{\omega}_i, \bar{\omega}_j \rangle]$ is positive definite.

The solutions of the Ramanujan vector field or its generalization (2.9) are also equipped with a metric which we call it again the Kähler metric. Let $\gamma(\tau)$ be a solution of R, where R is either the Ramanujan vector field or (2.9). The Hermitian form in the image of γ is defined to be

$$\partial_\tau \partial_{\bar{\tau}} \log \langle \alpha, \bar{\alpha} \rangle d\tau \otimes d\bar{\tau},$$

where α as in §2.2 or $\alpha = \alpha_1$ as in §2.3. In the next paragraphs we are going to compute this Hermitian form. Note that the corresponding metric is defined in the leaves of R and not in the moduli space T.

B.1 Poincaré metric

In the case of Elliptic curves the Kähler metric gives us the Poincaré metric in the upper half plane. Let $\mathbb{P}^1$ be the compactified moduli of elliptic curves parametrized with the j-invariant. It is with three marked points $0, 1, \infty$. We associate the ramification indices $2, 3, \infty$ to $0, 1, \infty$, respectively, and consider the uniformization $\mathbb{H} \to \mathbb{P}^1$ of $\mathbb{P}^1$ according to this indexing. For instance, the inverse of this uniformization can be defined through elliptic integrals:

$$j \mapsto \tau := \frac{\int_{\delta_1} \omega}{\int_{\delta_2} \omega},$$

where E_j is the elliptic curve with the j-invariant j as in (A.5), δ_i, $i = 1, 2$ generate $H_1(E_j, \mathbb{Z})$ with $\langle \delta_1, \delta_2 \rangle = -1$ and ω is a holomorphic 1-form on E_j. The pull-back of Kähler metric by this uniformization is the Poincaré metric in the upper half plane. This can be also seen in the framework of the moduli space T.

Recall our notations in §2.2. We define

$$\mathsf{G} : \mathsf{T} \to \mathrm{Mat}(2, \mathbb{C}), \tag{B.4}$$

$$\mathsf{G} := \begin{pmatrix} \langle \alpha, \bar{\alpha} \rangle & \langle \alpha, \bar{\omega} \rangle \\ \langle \omega, \bar{\alpha} \rangle & \langle \omega, \bar{\omega} \rangle \end{pmatrix} = \mathsf{P}^{\mathsf{tr}} \Psi^{-\mathsf{tr}} \overline{\mathsf{P}} = \begin{pmatrix} -2i\mathrm{Im}(x_1 \bar{x}_3) & x_3 \bar{x}_2 - x_1 \bar{x}_4 \\ x_4 \bar{x}_1 - x_2 \bar{x}_3 & -2i\mathrm{Im}(x_2 \bar{x}_4) \end{pmatrix},$$

where $\Psi := \begin{pmatrix} 0 & -1 \\ 1 & 0 \end{pmatrix}$. Let L be the locus of points $t \in \mathsf{T}$ such that the period matrix P in some basis δ_1, δ_2 of $H_1(E_t, \mathbb{Z})$ with $\langle \delta_1, \delta_2 \rangle = -1$ is of the form $\begin{pmatrix} \tau & -1 \\ 1 & 0 \end{pmatrix}$, $\tau \in \mathbb{H}$. We restricted G to this loci and we have

$$\mathsf{G}\,|_L = \begin{pmatrix} \tau & 1 \\ -1 & 0 \end{pmatrix} \begin{pmatrix} 0 & -1 \\ 1 & 0 \end{pmatrix} \begin{pmatrix} \bar{\tau} & -1 \\ 1 & 0 \end{pmatrix} = \begin{pmatrix} \bar{\tau}-\tau & -1 \\ 1 & 0 \end{pmatrix}.$$

We only need the $(1,1)$-entry of this matrix in order to see that Kähler metric is actually the Poincaré metric in the upper half plane:

$$\partial_\tau \partial_{\bar{\tau}} \log(\bar{\tau}-\tau) d\tau \otimes d\bar{\tau} = \frac{d\tau \otimes d\bar{\tau}}{(\bar{\tau}-\tau)^2}.$$

B.2 Kähler metric for moduli of mirror quintics

We repeat the same argument as of the previous paragraph for the compactified moduli of mirror quintics. This is $\mathbb{P}^1$ minus $\{0,1,\infty\}$. In this case we consider the ramification index $\infty,\infty,5$ to $z=0,1,\infty$, respectively, and get the monodromy covering $\mathbb{H} \to \mathbb{P}^1$ which we discussed in §4.6. Let $\tau_i,\ i=0,1,2,3$ be the meromorphic functions on $\mathbb{H}$ defined in §4.6. The Kähler metric in this case is not the Poincaré metric on $\mathbb{H}$. We define the map

$$\mathsf{G} : \mathsf{T} \to \mathrm{Mat}(4,\mathbb{C}),\ \ \mathsf{G}(t) := [\langle \alpha_i, \bar{\alpha}_j \rangle].$$

Let $L \subset \mathsf{T}$ be the one-dimensional locus in §4.2. We have

$$\mathsf{G}\,|_L = \begin{pmatrix} (\tau_0\bar{\tau}_1 - \tau_1\bar{\tau}_0 - \tau_2 + \bar{\tau}_2) & (\tau_0\bar{\tau}_3 - \tau_1 - \bar{\tau}_0\tau_3 + \bar{\tau}_1) & (\tau_0 - \bar{\tau}_0) & 1 \\ (\tau_0\tau_3 - \tau_1 - \tau_3\bar{\tau}_0 + \bar{\tau}_1) & -(\tau_3 - \bar{\tau}_3) & 1 & 0 \\ (\tau_0 - \bar{\tau}_0) & -1 & 0 & 0 \\ -1 & 0 & 0 & 0 \end{pmatrix}. \tag{B.5}$$

Therefore, the Hermitian form in $\mathbb{H}$ is given by

$$\partial_{\tau_0} \partial_{\bar{\tau}_0} (\tau_0\bar{\tau}_1 - \tau_1\bar{\tau}_0 - \tau_2 + \bar{\tau}_2) d\tau_0 \otimes d\bar{\tau}_0.$$

Using (B.5) we get the following format of the Hermitian form:

$$\left(\frac{\mathrm{Im}(\tau_3)}{\mathrm{Im}(\tau_1\bar{\tau}_0 + \tau_2)} + \frac{|\tau_0\tau_3 - \tau_1 - \tau_3\bar{\tau}_0 + \bar{\tau}_1|^2}{4\mathrm{Im}(\tau_1\bar{\tau}_0 + \tau_2)^2} \right) d\tau_0 \otimes d\bar{\tau}_0.$$

For the computer codes used in the computations of this section see the author's webpage [Supp Item 57]. The asymptotic behavior of the above metric near $z = 0,1,\infty$ have been obtained in Table 2, page 50 in [CdlOGP91]. In [Ali14] we can also find explicit expressions for such Kähler hermitian forms using different notations.

One of the most significant applications of metrics in Number Theory is in the proof of convergence of Poincaré series. For instance, the Eisenstein series are first defined in the following way:

$$E_{2i} := \sum_{0 \neq (m,n) \in \mathbb{Z}^2} \frac{1}{(n+m\tau)^{2i}}, \quad i \geq 2, \quad \tau \in \mathbb{H}. \tag{B.6}$$

One can compute the q-expansion of these series and one gets $2\zeta(2k)$ times those defined in (2.21), see for instance [Kob93, Zag08]. The convergence of the above series can be deduced from the fact that the quotient of the upper half plane by $\mathrm{SL}(2,\mathbb{Z})$ has finite volume with respect to the Poincaré metric, see [Kob93, Fre83]. In the present text we have generalized the Eisenstein series for mirror quintics, however, it is not at all clear how one can formulate series similar to (B.6) in our framework.

Appendix C
Integrality properties of Calabi-Yau modular forms

HOSSEIN MOVASATI, KHOSRO M. SHOKRI

We study the integrality properties of the coefficients of the mirror map attached to the generalized hypergeometric function ${}_nF_{n-1}$ with rational parameters and with a maximal unipotent monodromy. We present a conjecture on the p-integrality of the mirror map which can be verified experimentally. We prove its consequence on the N-integrality of the mirror map for the particular cases $1 \leq n \leq 4$. For $n = 2$ we obtain the Takeuchi's classification of arithmetic triangle groups with a cusp, and for $n = 4$ we prove that 14 examples of hypergeometric Calabi-Yau equations are the full classification of hypergeometric mirror maps with integral coefficients. As a by-product we get the integrality of the corresponding algebra of differential Calabi-Yau modular forms. These are natural generalizations of the algebra of classical modular and quasi-modular forms in the case $n = 2$.

C.1 Introduction

The integrality of the coefficients of the mirror map is a central problem in the arithmetic of Calabi-Yau varieties and it has been investigated in many recent articles [LY96a, LY96b, LY98, HLY96, Zud02, Sam09, KSVv3, KRNT]. The central tool in all these works has been the so-called Dwork method developed in [Dwo69, Dwo73]. It seems to us that the full consequences of Dwork method has not been explored, that is, to classify all hypergeometric differential equations with a maximal unipotent monodromy whose mirror maps have integral coefficients. In this appendix, we fill this gap and give a computable condition on the parameters of a hypergeometric function which conjecturally computes all the primes which appear in the denominators of the coefficients of the mirror map. We verify this conjecture and some of its consequences in many particular cases and give many computational evidence for its validness.

Let a_i, $i = 1, 2, \ldots, n$ be rational numbers, $0 < a_i < 1$, $a = (a_1, a_2, \ldots, a_n)$ and

$$F(a|z) := {}_nF_{n-1}(a_1,\dots,a_n;1,1,\dots,1|z) = \sum_{k=0}^{\infty} \frac{(a_1)_k\cdots(a_n)_k}{k!^n} z^k, \quad |z|<1$$

be the holomorphic solution of the generalized hypergeometric differential equation

$$\theta^n - z(\theta+a_1)(\theta+a_2)\cdots(\theta+a_n) = 0 \tag{C.1}$$

where $(a_i)_k = a_i(a_i+1)(a_i+2)\cdots(a_i+k-1)$, $(a_i)_0 = 1$, is the Pochhammer symbol and $\theta = z\frac{d}{dz}$. The first logarithmic solution in the Frobenius basis around $z=0$ has the form $G(a|z)+F(a|z)\log z$, where

$$G(a|z) = \sum_{k=1}^{\infty} \frac{(a_1)_k\cdots(a_n)_k}{(k!)^n}\Big[\sum_{j=1}^{n}\sum_{i=0}^{k-1}\Big(\frac{1}{a_j+i}-\frac{1}{1+i}\Big)\Big]z^k. \tag{C.2}$$

The mirror map

$$q(a|z) =: z\exp\Big(\frac{G(a|z)}{F(a|z)}\Big)$$

is a natural generalization of the Schwarz function.

For a prime p and a formal power series $f \in \mathbb{Q}[[z]]$ with rational coefficients we say that f is p-integral if p does not appear in the denominator of its coefficients or equivalently f induces a formal power series in $\mathbb{Z}_p[[z]]$. For a rational number x such that p does not divide the denominator of x, we define

$$\delta_p(x) := \frac{x+x_0}{p},$$

where $0 \le x_0 \le p-1$ is the unique integer such that p does not divide the denominator of $\delta_p(x)$. We call δ_p the Dwork operator. Throughout the text we assume all parameters a_i's in the hypergeometric equation (C.1) are p-integers and such p is called a good prime.

Conjecture 1 *Let $q(a|z)$ be the mirror map of the generalized hypergeometric function, defined as above. Then $q(a|z)$ is p-integral if and only if*

$$\{\delta_p(a_1),\delta_p(a_2)\} = \{a_1,a_2\},\ \textit{or}\ \{1-a_1,1-a_2\}, \quad \textit{for}\ n=2 \tag{C.3}$$

and

$$\{\delta_p(a_1),\delta_p(a_2),\delta_p(a_3),\dots,\delta_p(a_n)\} = \{a_1,a_2,a_3,\dots,a_n\}, \quad \textit{for}\ n\neq 2. \tag{C.4}$$

The conjecture for $n=1$ is an easy exercise. Due to the Euler identity for the Gauss hypergeometric function, the case $n=2$ appears different from other cases, see §C.7. For general n we prove the only if part and give computational evidence for the validity of the other direction, see §C.10. In the literature one is mainly interested to classify all N-integral mirror maps. This means that there is a natural number N such that $q(a|Nz)$ has integral coefficients.

Remark. One can see that the order of bad primes in the coefficients of F, G and consequently in q has at most polynomial growth. Hence there is a number N_p such that, $q(a|N_p z)$ is p-integral. Therefore p integrality of $q(a|z)$ for almost all p implies that $q(a|z)$ is N-integral.

The Conjecture 1 easily implies

Conjecture 2 *The mirror map $q(a|z)$ is N-integral if and only if for any good prime p we have (C.3) for $n=2$ and (C.4) for $n \geq 3$.*

Theorem 13 *We have*

1. *For an arbitrary n the only if part of Conjecture 1 and Conjecture 2 are true.*
2. *Conjecture 2 for $n = 1,2,3,4$ is true.*

The first part of the above theorem together with an exact formula for the smallest number N such that $q(a|Nz)$ is integral, was a conjecture in the context of mirror symmetry. The case $a_i = \frac{i}{n+1}$, $i = 1,2,\ldots,n$, where $n+1$ is a prime number (resp. power of a prime number) is proved by Lian-Yau in [LY96a] (resp. by Zudilin in [Zud02]). A general format is formulated by Zudilin in [Zud02] and is proved by Krattenthaler-Rivoal in [KRNT]. According to Conjecture 2 the N-integrality of the mirror map implies that for a fixed good prime p the map δ_p acts on the set $\{a_1,\ldots,a_n\}$ as a permutation. This set is decomposed into subsets of cardinality $\phi(m_i)$, $m_i > 1$, $i = 1,2,\ldots,k$, where ϕ is the arithmetic Euler function. This is done according to whether two elements have the same denominator or not. The numbers m_i, $i = 1,2,\ldots$ are all possible denominators in the set $\{a_1,\ldots,a_n\}$. Such a decomposition is invariant under the permutation induced by δ_p. We conclude that the number of N-integral mirror maps is equal to the number of decompositions:

$$n = \phi(m_1) + \cdots + \phi(m_k). \tag{C.5}$$

Like in the classical case of the partition of a number with natural numbers we have the following generating function:

$$24x^2 - 1 + \prod_{m=2}^{\infty} \frac{1}{1 - x^{\phi(m)}} = x + 28x^2 + 4x^3 + 14x^4 + 14x^5 + 40x^6 + 40x^7 + \cdots$$

(for the coefficient of x^2 see Table 1). Conjecture 2 says that the coefficient of x^n counts the number of N-integral mirror maps with n-parameters $a_1, a_2, \ldots, a_n$. An immediate consequence of (C.5) is that the number of N-integral mirror maps for $n = 2\ell + 1$, $\ell \geq 2$ exactly equals with $n = 2\ell$ ones. Since $\phi(m)$, for $m > 2$ is even, for $n = 2\ell + 1$ in the right hand side of (C.5), one of the m_i's is 2 or equivalently one of a_i is $\frac{1}{2}$. After canceling this from both sides we return to the case $n = 2\ell$. Therefore, the number of N-integral mirror maps for $n = 2\ell$ and $2\ell + 1$ are in a one to one correspondence.

For any fixed n, we can use conjecture 2 and classify all N-integral mirror maps. For instance for $n = 2$, the mirror map is N-integral if and only if $\{a_1, a_2\}$ belongs to the list in Table 1. The first four cases correspond to $\{\delta_p(a_1), \delta_p(a_2)\} = \{a_1, a_2\}$ and the others correspond to $\{\delta_p(a_1), \delta_p(a_2)\} = \{1-a_1, 1-a_2\}$. If we set $\{a_1, a_2\} = \{\frac{1}{2}(1 \pm \frac{1}{m_1} - \frac{1}{m_2})\}$ with $m_1, m_2 \in \mathbb{N}$ and $\frac{1}{m_1} + \frac{1}{m_2} > 1$ then the monodromy group of the Gauss hypergeometric equation (C.1) is a triangle group of type (m_1, m_2, ∞). In this case the above classification is reduced to the Takeuchi's classification in [Tak77] of arithmetic triangle groups with a cusp, see [DGMS13].

For $n \geq 4$ even and with an N-integral mirror map $q(a|z)$, the set $\{a_1, a_2, \dots, a_n\}$ is conjecturally invariant under $x \mapsto 1-x$, see §C.2 and so we may identify $a = (a_1, a_2, \dots, a_n)$ with its $\frac{n}{2}$ elements in the interval $[0, \frac{1}{2}]$. Below we just list these elements. The case $n = 4$ has many applications in mirror symmetry. We find that $q(a|z)$ is N-integral if and only if (a_1, a_2, a_3, a_4) belongs to the well-known 14 hypergeometric cases of Calabi-Yau equations. The first two elements (a_1, a_2) are given in Table 1. Note that in [DM06], these 14 cases are classified through properties of the monodromy group of the differential equation (C.1) which comes from the variation of Hodge structures of Calabi-Yau varieties. The fact that the corresponding mirror maps are N-integral and are the only ones with this property is a non-trivial statement. For $n = 6$ we find 40 examples of N-integral mirror maps. In this case (a_1, a_2, a_3) are given in Table 1. Finding 40 examples of one parameter families of Calabi-Yau varieties of dimension 5 may be done in a similar way as in the $n = 4$ case. This is left for a future work. Note that for an arbitrary n the Picard-Fuchs equation of the Dwork family $x_1^{n+1} + \dots + x_{n+1}^{n+1} - (n+1)z^{-\frac{1}{n+1}} x_1 x_2 \cdots x_{n+1} = 0$ corresponds to the case $a_i = \frac{i}{n+1}, \ i = 1, 2, \dots, n$.

The discussion for $n = 2$ suggests that the N-integrality of the mirror map is related to the arithmeticity of the monodromy group of (C.1). However, note that for $n = 4$ it is proved that the seven of the fourteen examples are thin, that is, they have infinite index in $\mathrm{Sp}(4, \mathbb{Z})$, see [BT14]. Note also that, we know a complete classification of the Zariski closure of the monodromy group of (C.1), see [BH89]. Since in the present text we have constructed a new kind of modular forms theory for these examples of thin groups, see also §C.11, it is reasonable to weaken the notion of arithmeticity so that the monodromy group of (C.1) with an N-integral mirror map becomes arithmetic in this new sense. For a discussion of thin groups and their applications see Sarnak's lecture notes [Sar12] and the references therein. Finding the smallest N is not a trivial problem and is discussed in the articles [LY96a, Zud02, KRNT] for particular cases including the 14 cases in Table 1. For the following corollary we assume this N.

Corollary 1 *Let $n = 4$ and consider one of the 14 cases in Table 1. Let also $z(q)$ be the inverse of the mirror map $q(z) := \frac{1}{N} q(a|Nz)$ and*

$$u_0(z) := z, \quad u_i(z) := \theta^i(F(Nz)), \ i = 0, 1, 2, 3,$$
$$u_{i+4}(z) := F(Nz)\theta^i(G(Nz)) - G(Nz)\theta^i(F(Nz)), \quad i = 1, 2.$$

We have

$$u_0(q), u_i(z(q)) \in \mathbb{Z}[[q]], \ i = 0, 1, \dots, 6. \tag{C.6}$$

The field of differential Calabi-Yau modular forms is generated by (C.6) and it is invariant under $q\frac{\partial}{\partial q}$. There are many analogies between $\mathbb{Q}(u_i(z(q)), \ i = 0, 1, \dots, 6)$

and the field of quasi-modular forms. This includes functional equations with respect to the monodromy group of (C.1), the corresponding Halphen equation and so on. However, note that the former field is of transcendental degree 3, whereas this new field is of transcendental degree 7. For $n = 2$ a similar discussion as in Corollary 1 leads us to the theory of (quasi) modular forms for the monodromy group of (C.1), see for instance [DGMS13]. F. Beukers informed us about the article [Roq14] in which the author proves Conjecture 3 in §C.6 using results in differential Galois theory. As a consequence the if part of the Conjecture 2 follows easily. Our proof of the Conjecture 2 for $n = 1,2,3,4$ is elementary and self-content.

$n=2$
$(1/2,1/2),(2/3,1/3),(3/4,1/4),(5/6,1/6),$ $(1/6,1/6),(1/3,1/6),(1/2,1/6),(1/3,1/3),(2/3,2/3),$ $(1/4,1/4),(1/2,1/4),(3/4,1/2),(3/4,3/4),(1/2,1/3),$ $(2/3,1/6),(2/3,1/2),(5/6,1/3),(5/6,1/2),(5/6,2/3),$ $(5/6,5/6),(3/8,1/8),(5/8,1/8),(7/8,3/8),(7/8,5/8),$ $(5/12,1/12),(7/12,1/12),(11/12,5/12),(11/12,7/12)$
$n=4$
$(1/2,1/2),(1/3,2/3),(1/4,1/2),(1/4,1/4),(2/5,1/5),$ $(3/8,1/8),(3/10,1/10),(1/2,1/6),(1/2,1/3),(1/3,1/6),$ $(1/6,1/6),(1/3,1/4),(1/4,1/6),(5/12,1/12)$
$n=6$
$(1/2,1/2,1/2),(1/3,1/3,1/3),(1/2,1/2,1/4),(1/2,1/4,1/4),$ $(1/4,1/4,1/4),(1/2,1/2,1/3),(1/2,1/3,1/3),(1/2,1/2,1/6),$ $(1/2,1/3,1/6),(1/3,1/3,1/6),(1/2,1/6,1/6),(1/3,1/6,1/6),$ $(1/6,1/6,1/6),(3/7,2/7,1/7),(1/2,3/8,1/8),(3/8,1/4,1/8),$ $(4/9,2/9,1/9),(1/2,2/5,1/5),(1/2,3/10,1/10)(1/2,1/3,1/4),$ $(1/3,1/3,1/4),(1/3,1/4,1/4),(1/2,1/4,1/6),(1/3,1/4,1/6),$ $(1/4,1/4,1/6),(1/4,1/6,1/6),(1/2,5/12,1/12),(5/12,1/3,1/12),$ $(5/12,1/4,1/12),(5/12,1/6,1/12),(5/14,3/14,1/14),(2/5,1/3,1/5),$ $(7/18,5/18,1/18),(2/5,1/4,1/5),(3/10,1/4,1/10),(3/8,1/3,1/8),$ $(3/8,1/6,1/8),(2/5,1/5,1/6),(1/3,3/10,1/10),(3/10,1/6,1/10)$

Table 1: N-integral hypergeometric mirror maps.

For the computer codes used in the present section see the first author's webpage [Supp Item 58].

C.2 Dwork map

As far as we know, the main argument in the literature for proving the p- or N-integrality of mirror maps is the work of B. Dwork in [Dwo73, Dwo69].

The Dwork map is defined in the following way:

$$\delta_p : \mathbb{Z}_p \to \mathbb{Z}_p, \ \sum_{s=0}^{\infty} x_s p^s \longmapsto 1 + \sum_{s=0}^{\infty} x_{s+1} p^s, \quad 0 \le x_s \le p-1$$

(and so $p\delta_p(x) - x = p - x_0$ for $x \in \mathbb{Z}_p$). Let $\tilde{\mathbb{Z}}_p$ be the set of p-integral rational numbers. We have a natural embedding $\tilde{\mathbb{Z}}_p \hookrightarrow \mathbb{Z}_p$. The map δ_p leaves $\tilde{\mathbb{Z}}_p$ invariant because for $x \in \tilde{\mathbb{Z}}_p$, $\delta_p(x)$ is the unique number such that $p\delta_p(x) - x \in \mathbb{Z} \cap [0, p-1]$. For a rational number $x = \frac{x_1}{x_2}$, $x_1, x_2 \in \mathbb{Z}$ and a prime p which does not divide x_2, we have

$$\delta_p(x) := \frac{p^{-1} x_1 \bmod x_2}{x_2}, \tag{C.7}$$

where p^{-1} is the inverse of p mod x_2 and x_1 and x_2 may have common factors. The denominators of x and $\delta_p(x)$ are the same and $\delta_p(1-x) = 1 - \delta_p(x)$. For any finite set of rational numbers, there is a finite decomposition of prime numbers such that in each class the function δ_p is independent of the prime p.

The following proposition easily follows from (C.7). It will play an important role in the proof of Theorem 13.

Proposition 32 *Let $0 < x = \frac{t}{sq^y} < 1$ be a rational number, where q is a prime and $(q,s) = 1$ and t and sq^y may have common factors. We have*

1. *For primes p such that $p^{-1} \equiv -1 \pmod{sq^y}$ we have $\delta_p(x) = 1 - x$.*
2. *If $y = 0$, that is the denominator of x is not divisible by q, then for primes p such $p^{-1} \equiv q \pmod{s}$ we have*

$$\delta_p(x) = qx - i, \ \textit{for some } i = 0, 1, 2, \ldots, q-1.$$

3. *For primes $p^{-1} \equiv s + q \pmod{sq^y}$ we have*

$$\delta_p(x) = qx + \frac{r}{q^y} - i, \ \textit{for some } \ r = 0, 1, \ldots, q^y - 1, \ i = 0, 1, \ldots, q.$$

4. *If $y \ge 1$, that is, the denominator of x may be divisible by q, then for any $0 \le m \le y$ and prime p with $p^{-1} \equiv 1 + q^{y-m} s \pmod{sq^y}$ either we have $\delta_p(x) = x$ or we have*

$$\delta_p(x) = x + \frac{r}{q^m} - i, \ \textit{for some } \ r = 1, \ldots, q^m - 1, \ i = 0, 1.$$

C.3 Dwork lemma and theorem on hypergeometric functions

In this section we mention some lemmas and a theorem of Dwork. The following lemma is crucial in the argument of N-integrality of the mirror map. Since the non-trivial application of this lemma was first given by Dwork, it is known as Dwork lemma, however, Dwork himself in [DGS94] associates it to Dieudonné.

Lemma 1 *Let $f(z) \in 1 + z\mathbb{Q}_p[[z]]$. Then $f(z) \in 1 + z\mathbb{Z}_p[[z]]$, if and only if*

$$\frac{f(z^p)}{(f(z))^p} \in 1 + p\mathbb{Z}_p[[z]].$$

For a more general statement and the proof see [DGS94], p. 54. For the following theorem the reader is referred to [Dwo69, Dwo73].

Theorem 14 *Let F, G as before. We have*

$$\frac{G(\delta_p(a)|z^p)}{F(\delta_p(a)|z^p)} \equiv p\frac{G(a|z)}{F(a|z)} \pmod{p\mathbb{Z}_p[[z]]}.$$

C.4 Consequences of Dwork's theorem

The mirror map is the invertible function $q(a|z) = z\exp(\frac{G(a|z)}{F(a|z)})$. The inverse of the mirror map is important for the construction of differential Calabi-Yau modular forms. In this section we give conditions for integrality of the mirror map.

Lemma 2 *Let $a_1, a_2, \ldots, a_n \in \mathbb{Q}$ with the conditions* (C.3) *and* (C.4) *for a prime p. Then*

$$q(a|z) \in z\mathbb{Z}_p[[z]].$$

Proof. Let $f(z) = \frac{G(a|z)}{F(a|z)}$. Since $f(z) \in z\mathbb{Q}[[z]]$, so $\exp(f(z)) \in 1 + z\mathbb{Q}[[z]]$. Our assumption and Theorem 14 imply that

$$f(z^p) - pf(z) = p \cdot g(z), \quad g(z) \in z\mathbb{Z}_p[[z]].$$

Since $\exp(p.g(z)) = 1 + \sum_{k=1}^{\infty} \frac{p^k}{k!} g(z)^k$ and $ord_p(k!) < k$, we find that

$$\frac{\exp f(z^p)}{(\exp f(z))^p} = \exp(p.g(z)) \in 1 + pz\mathbb{Z}_p[[z]].$$

Now by applying Lemma 1, the result follows.

Lemma 3 *If $q(a|z)$ is p-integral then*

$$\frac{G(\delta_p(a)|z)}{F(\delta_p(a)|z)} \equiv \frac{G(a|z)}{F(a|z)} \pmod{p\mathbb{Z}_p[[z]]}. \tag{C.8}$$

Furthermore if $q(a|z)$ is p-integral for all except a finite number of primes then the above congruence is an equality for all good primes p.

We conjecture that in Lemma 3, the congruence (C.8) is an equality. This together with Conjecture 3 in §C.6 imply Conjecture 1.

Proof. Let

$$f(z) := \frac{G(a|z)}{F(a|z)}, \quad f'(z) := \frac{G(\delta_p(a)|z)}{F(\delta_p(a)|z)}.$$

By Lemma 1 $\exp(f)$ is p-integral if and only if $\exp(f(z^p) - pf(z)) \in 1 + p\mathbb{Z}_p[[z]]$ and by Theorem 14 $f'(z^p) \equiv pf(z) \pmod{p\mathbb{Z}_p[[z]]}$. Combining these two facts

$$\exp(f(z^p) - f'(z^p)) \in 1 + pz\mathbb{Z}_p[[z]],$$

or $f(z^p) - f'(z^p) = \log(1 + pzg(z))$, for some $g(z) \in \mathbb{Z}_p[[z]]$. But

$$\log(1 + pzg(z)) = \sum_{n=1}^{\infty} (-1)^n \frac{p^n z^n g(z)^n}{n} \in pz\mathbb{Z}_p[[z]].$$

Now, let us prove the second part. Since the number of bad primes is finite, our hypothesis implies that $q(a|z)$ is p-integral for all good primes except a finite number which may includes bad primes. Let c is the common factor of the denominators of a_i's and let r be a prime residue of c. From (C.7) for $p \equiv r \pmod{c}$ the value of $\delta_p(a_i)$ is independent of the special member of this class of primes. On the other hand by assumption congruency (C.8) holds for almost all primes of this class and so it must be equality. Now running over all prime residues r gives the result.

Remark 17 If the congruence (C.8) does not happen then by Lemma 3, $q(a|z)$ is not p-integral. In fact this turns out to be a fast way to check the non p-integrality than checking the non p-integrality of $q(a|z)$ directly. For instance, for $n = 2$, $p = 101, a_1 = 169/330,\ \ a_2 = 139/330$ the truncated $q(a|z) \mod z^k$, $k < 101$ is p-integral but $q(a|z)$ is not p-integral because the congruency (C.8) fails at the power z^2.

Lemma 4 *If the mirror map $q(a|z)$ is p-integral then $q(\delta_p(a)|z)$ is also p-integral.*

Proof. The proof follows directly from the congruency (C.8). Note that, if $f \in p\mathbb{Z}_p[[z]]$, then $e^f \in \mathbb{Z}_p[[z]]$.

C.5 Proof of Theorem 13, Part 1

The only if part of both conjectures 1 and 2 follows from Lemma 2. So, it remains to prove Conjecture 2 for $n = 1, 2, 3, 4$.

C.6 A problem in computational commutative algebra

The only missing step for the proof of Conjecture 2 is a solution to the following conjecture:

Conjecture 3 *Let $n \neq 2$ and $a_i, b_i \in \mathbb{Q},\ \ i = 1, 2, \ldots, n$ with $0 < a_i, b_i < 1$, we have the equality of formal power series*

$$\frac{G(b_1,b_2,\ldots,b_n|z)}{F(b_1,b_2,\ldots,b_n|z)} = \frac{G(a_1,a_2,\ldots,a_n|z)}{F(a_1,a_2,\ldots,a_n|z)}, \tag{C.9}$$

if and only if

$$\{b_1,b_2,\ldots,b_n\} = \{a_1,a_2,\ldots,a_n\}. \tag{C.10}$$

Let

$$\frac{G(a|z)}{F(a|z)} = \sum_{i=1}^{\infty} C_k(a)z^k, \quad C_k(a) \in \mathbb{Q}[a]$$

and $I_{n,m}$ be the ideal in $\mathbb{Q}[a,b]$ generated by $C_k(a)-C_k(b)$, $k=1,2,\ldots,m$ and $I_n = I_{n,\infty}$. The above conjecture over $\mathbb{C}$ is false, that is, the variety given by I_n has many components other than those obtained by the permutation (C.10). Therefore, the above conjecture is equivalent to say that such extra components have no non-trivial $\mathbb{Q}$-rational points. In this section we will try to avoid Conjecture 3 in such a general context. Instead, we will use the structure of the operator δ_p in order to reduce the number of variables and try to solve Conjecture 3 for such particular cases. Below, we are going to write down the primary decomposition of $I_{n,m}$ in the ring $\mathbb{Q}[a,b]$ for many particular cases. We have used the Gianni-Trager-Zacharias algorithm implemented in `Singular` under the command `primdecGTZ`, see [GPS01].

C.7 The case $n=2$

The case $n=2$ is different because of the Euler identity. We have the primary decomposition

$$\begin{aligned} I_{2,2} = {} & \langle a_2-b_2, a_1-b_1\rangle \cap \langle a_2-b_1, a_1-b_2\rangle \cap \\ & \langle a_2+b_2-1, a_1+b_1-1\rangle \cap \langle a_2+b_1-1, a_1+b_2-1\rangle \end{aligned} \tag{C.11}$$

in the ring $\mathbb{Q}[a,b]$. Therefore, the equality in Conjecture 3 is valid if and only if

$$\{b_1,b_2\} = \{a_1,a_2\} \text{ or } \{1-a_1, 1-a_2\}. \tag{C.12}$$

The second possibility for $\{b_1,b_2\}$ is due to the Euler identity for the Gauss hypergeometric function ${}_2F_1(a,b,c|z) = \sum_{k=0}^{\infty} \frac{(a)_k(b)_k}{k!(c)_k} z^k$:

$${}_2F_1(a,b,c|z) = (1-z)^{c-a-b}\,{}_2F_1(c-a,c-b,c|z).$$

Note that we put $c=1$ and the same equality is valid for the logarithmic solution. This proves Theorem 13, Part 2 for $n=2$.

C.8 The symmetry

Let $b_i = 1 - a_i,\ i = 1, 2, \ldots, n$ and let us restrict the ideal $I_{n,m}$ to this locus. For $n = 3$ we have the primary decomposition

$$I_{3,3} = \langle a_2 + a_3 - 1, 2a_1 - 1\rangle \cap \langle 2a_3 - 1, a_1 + a_2 - 1\rangle \cap \langle 2a_2 - 1, a_1 + a_3 - 1\rangle \cap$$
$$\langle a_3 - 1, a_2 - 1, a_1 - 1\rangle \cap \langle a_3, a_2, a_1\rangle$$

and for $n = 4$ we have the primary decomposition

$$I_{4,4} = \langle a_2 + a_3 - 1, a_1 + a_4 - 1\rangle \cap \langle a_3 + a_4 - 1, a_1 + a_2 - 1\rangle \cap \langle a_2 + a_4 - 1, a_1 + a_3 - 1\rangle \cap$$
$$\langle a_3^2 + a_4^2 - 2a_3 - 2a_4 + 2, a_2 + a_3 - 2, a_1 + a_4 - 2\rangle \cap \langle a_3^2 + a_4^2, a_2 + a_3, a_1 + a_4\rangle \cap$$
$$\langle a_3 + a_4, a_2^2 + a_4^2, a_1 + a_2\rangle \cap \langle a_3^2 + a_4^2 - 2a_3 - 2a_4 + 2, a_2 + a_4 - 2, a_1 + a_3 - 2\rangle \cap$$
$$\langle a_3^2 + a_4^2, a_2 + a_4, a_1 + a_3\rangle \cap \langle a_3 + a_4 - 2, a_2^2 + a_4^2 - 2a_2 - 2a_4 + 2, a_1 + a_2 - 2\rangle.$$

We conclude that Conjecture 3 is true for $n = 3, 4$ and $b_i = 1 - a_i$. Note that for $n = 4$ the components which are not in the variety $\{a_i,\ i = 1, 2, 3, 4\} = \{1 - a_i,\ i = 1, 2, 3, 4\}$ do not have $\mathbb{Q}$-rational points.

C.9 Proof of Theorem 13, Part 2

The only missing step is the verification of Conjecture 3. The case $n = 1$ can be done by hand and the case $n = 2$ is done in §C.7. For $n \geq 3$ we do not get the primary decomposition of $I_{n,n}$ in Singular, therefore, we do not have a proof for Conjecture 3 with arbitrary parameters a_i and b_i. However, we may try to prove it for particular classes of parameters a_i, b_i. The particular cases that appear in this section are motivated by the structure of δ_p described in Proposition 32.

First, let us take primes p such that $\delta_p(x) = 1 - x$, see Proposition 32, Item 1. In this case we have the primary decompositions in §C.8 which implies that the set of a_i's is invariant under $x \mapsto 1 - x$. For $n = 3$ we conclude that one of the parameters, let us say a_3, is $\frac{1}{2}$ and $a_1 = 1 - a_2$. Now, restricted to $a_3 = b_3 = \frac{1}{2}$ we have the primary decomposition:

$$I_{3,5} = \langle a_2 - b_1, a_1 - b_2\rangle \cap \langle a_2 - b_2, a_1 - b_1\rangle \cap \langle b_2 - 1, a_2 + b_1 - 1, a_1 - 1\rangle \cap$$
$$\langle b_2 - 1, a_2 - 1, a_1 + b_1 - 1\rangle \cap \langle b_1 - 1, a_2 + b_2 - 1, a_1 - 1\rangle \cap \langle b_1 - 1, a_2 - 1, a_1 + b_2 - 1\rangle.$$

This proves Conjecture 3 in this particular case. Note that we use the fact that none of parameters is 1.

For $n = 4$, in a similar way, we can assume that $a_3 = 1 - a_1$ and $a_4 = 1 - a_2$. Again we do not get the primary decomposition of $I_{4,4}$ restricted to the parameters $a_3 = 1 - a_1,\ a_4 = 1 - a_2, b_3 = 1 - b_1,\ b_4 = 1 - b_2$. We further use the structure of

δ_p in order to reduce the number of variables so that we can compute the primary decomposition of $I_{4,4}$.

For $q = 2$ or 3 fixed, let us consider the case in which q does not appear in the denominators of a_i's. Using Proposition 32, Item 2, we take primes p such that $\delta_p(a_1) = qa_1 - i$ and $\delta_p(a_2) = qa_2 - j$, where i, j are some integers between 0 and $q-1$.

Lemma 5 *For $q = 2,3$ and $i, j = 0,1,\dots,q-1$, Conjecture 3 is true for $n = 4$ and*

$$a_3 = 1 - a_1, a_4 = 1 - a_2,$$

$$b_1 = qa_1 - i, b_2 = qa_2 - j,\ b_3 = 1 - b_1,\ b_4 = 1 - b_2.$$

In each case the variety $V(I_{4,4})$ is a point with rational coordinates and if we take $0 < a_i \le \frac{1}{2}$ then we get 7 of 14-hypergeometric cases.

Proof. The proof is purely computational. For $q = 2$ or 3 we have considered q^2 cases. In each case, we find that the variety $V(I_{4,4})$ consists only of one point. It has rational coordinates. If we assume that $0 < a_i \le \frac{1}{2}$ then for $q = 2$ we have only the solutions $(a_1, a_2) = (\frac{1}{5}, \frac{2}{5}),\ (\frac{1}{3}, \frac{1}{3})$. For $q = 3$ we get

$$(a_1, a_2) = (\frac{1}{2}, \frac{1}{2}), (\frac{1}{4}, \frac{1}{2}), (\frac{1}{4}, \frac{1}{4}), (\frac{1}{5}, \frac{2}{5}), (\frac{1}{8}, \frac{3}{8}), (\frac{1}{10}, \frac{3}{10}).$$

Now, let us assume that 2 and 3 appears in the denominators of a_i's. We are going to use Proposition 32, Item 4 for $q = 2,3$ and $m = 1$. For $q = 2$, we take primes p such that $\delta_p(a_i) = a_i + r_i,\ i = 1,2$, where $\{r_1, r_2\} \subset \{0, \frac{1}{2}, -\frac{1}{2}\}$. In a similar way, for $q = 3$ we take primes p such that $\delta_p(a_i) = a_i + s_i,\ i = 1,2$, where $\{s_1,\ s_2\} \subset \{0, \frac{1}{3}, -\frac{2}{3}, \frac{2}{3}, -\frac{1}{3}\}$. We can assume that r_1 and r_2 (resp. s_1 and s_2) are not simultaneously zero. Let $I_{4,4,2,r_1,r_2}$, be $I_{4,4}$ restricted to the parameters:

$$a_3 = 1 - a_1, a_4 = 1 - a_2,$$

$$b_1 = a_1 + r_1, b_2 = a_2 + r_2,\ b_3 = 1 - b_1,\ b_4 = 1 - b_2,$$

and in a similar way define $I_{4,4,3,s_1,s_2}$. The variety $V(I_{4,4,2,r_1,r_2}) \cap V(I_{4,4,3,s_1,s_2})$ is just one point. Moreover, it has rational coordinates. If we put the condition $0 < a_i \le \frac{1}{2}$ we get the remaining 7 examples in the list (C.1).

C.10 Computational evidence for Conjecture 1

For $n = 2$ and $\{a_1, a_2\} = \{\frac{1}{2}(1 \pm \frac{1}{m_1} - \frac{1}{m_2})\}$ with $m_1, m_2 \in \mathbb{N}$ and $\frac{1}{m_1} + \frac{1}{m_2} > 1$ the monodromy group of (C.1) is a triangle group of type (m_1, m_2, ∞) and we have checked Conjecture 1 for the truncated $q(a|z) \pmod{q^{182}}$ and for all the cases

$$m_1 \le m_2 \le 24, \text{ or } m_1 \le 24,\ m_2 = \infty$$

and primes $2 \leq p \leq 181$. For Conjecture 3 (which implies Conjecture 2), apart from its verification in particular cases done in §C.6, we have checked it for many other special loci in the parameter space a, b. The strategy is always to reduce the number of variables so that we can compute the primary decomposition of $I_{n,m}$ by a computer. The detailed discussion of this topic will be written somewhere else.

C.11 Proof of Corollary 1

Corollary 1 follows from Theorem 13, Part 2 for $n = 4$ in the case of good primes and for bad primes follows from Theorem 1 in [KRNT]. Indeed the kth coefficient of F— the holomorphic solution of any equation in Table 1 for $n = 4$— is a product of expressions like

$$A_s(k) := \frac{(r_1/s)_k \cdots (r_m/s)_k}{k!^m}$$

where $r_1, \cdots, r_m$ is a complete set of prime residues of s. Hence $m = \phi(s) = s(1-\frac{1}{p_1})\cdots(1-\frac{1}{p_l})$, where $s = p_1^{\nu_1}\cdots p_l^{\nu_l}$, and one can easily check that $N_s^k A_s(k) \in \mathbb{Z}$, where $N_s = s^m \prod_{p|s} p^{m/p-1}$ (see [Zud02], Lemma 1). The number $N = \prod_s N_s$ makes $F(Nz)$ with integer coefficients. Since the ring $\mathbb{Z}[[z]]$ is closed under the operator θ, so $u_i(z) \in \mathbb{Z}[[z]]$ for $i = 0, \ldots, 4$. For u_5 we proceed as follows. Since $u_1(z) \in 1 + z\mathbb{Z}[[z]]$, it is enough to prove the statement for $W(a|z) := \frac{u_5(z)}{u_1(z)^2} = \theta(\frac{G}{F})$ (here we need to emphasize that u_i's depend on a). First let p be a good prime. Acting θ on the both sides of the congruence in Theorem 14 we get $W(\delta_p(a)|z^p) - W(a|z) \in \mathbb{Z}_p[[z]]$. Since for $k = 1, \ldots, p-1$, the kth coefficients of F and G are p-integral, with a inductive argument we conclude that $W(a|z) \in \mathbb{Z}_p[[z]]$. Now for bad primes we use Theorem 1 in [KRNT], saying that $q(z)$ has integer coefficients. Hence applying Lemma 1 for $q(z)$, implies Theorem 14 for bad primes, with convention $\delta_p(a) = a$ and replacing (z, z^p) by (Nz, Nz^p) in the statement. Then we can repeat the above argument to get integrality of $W(a|z)$. Now since $q(z)$ is integral, so converting all u_i's in q-coordinate remain them with integer coefficients.

The most important modular object arising from the periods of Calabi-Yau varieties is the Yukawa coupling:

$$\mathsf{Y} := n_0 \frac{u_1^4}{(u_5 + u_1^2)^3(1-z)} = n_0 + \sum_{d=1}^{\infty} n_d d^3 \frac{q^d}{1-q^d}.$$

Here, $n_0 := \int_M \omega^3$, where M is the A-model Calabi-Yau threefold of mirror symmetry and ω is the Kähler form (the Picard-Fuchs equation (C.1) is satisfied by the periods of B-model Calabi-Yau threefold). The numbers n_d are supposed to count the number of rational curves of degree d in a generic M. For the case $a_i = \frac{i}{5}$, $i = 1, 2, 3, 4$ few coefficients n_d are given by $n_d = 5, 2875, 609250, 317206375, \ldots$.

Appendix D
Variations of Hodge structures, Lyapunov exponents and Kontsevich's formula

CARLOS MATHEUS

In 1996, Kontsevich [Kon96] discovered a beautiful formula relating the features of certain *variations of Hodge structures* of weight 1 and some dynamical quantities called *Lyapunov exponents* in the context of a natural $SL(2,\mathbb{R})$-action on the moduli spaces of Abelian differentials on Riemann surfaces (a subject nowadays known as *Teichmüller dynamics*). A detailed proof of Kontsevich's formula was given in 2001 by Forni [Fo01]. More recently, a long work (of 127 pages) of Eskin-Kontsevich-Zorich [EKZ14] shows that an analytic version of Riemann-Roch-Hirzebruch-Grothendieck theorem can be used to reinterpret Kontsevich's formula in terms of flat geometry of Abelian differentials. Also, Chen [C13] and Chen-Möller [CM12] related Kontsevich's formula to intersection theory and slopes of divisors in moduli spaces.

Besides their intrinsic interest, the circles of ideas developed to justify Kontsevich's formula in a rigorous way were recently employed to answer some questions with eclectic motivations (from Statistical Mechanics, Algebraic Geometry, etc.). For example:

- P. Ehrenfest and T. Ehrenfest introduced in 1912 a model of Lorenz gases called *wind-tree model*. The physicists Hardy and Weber studied in 1980 the $\mathbb{Z}^2$-periodic version of Ehrenfest's wind-tree model (where a billiard with usual reflection laws on the plane $\mathbb{R}^2$ is constructed by placing rectangular obstacles of dimensions $a \times b$, $0 < a, b < 1$, are centered at each element of $\mathbb{Z}^2$). Hardy-Weber conjectured an *abnormal* rate of diffusion of typical trajectories (in the sense that almost all orbits of the billiard would diffuse faster than the prediction of the Central Limit Theorem for chaotic billiards). By exploiting the ideas around Kontsevich's formula, Delecroix-Hubert-Lelièvre [DHL14] confirmed Hardy-Weber conjecture in a strong form: they proved that the rate of diffusion is given by a certain Lyapunov exponent in Teichmüller dynamics and the precise value $2/3$ of this Lyapunov exponent can be computed essentially from Kontsevich's formula; in particular, the rate of diffusion is abnormal because $2/3 > 1/2$ and it is *independent* of the dimensions of the obstacles in the $\mathbb{Z}^2$-periodic wind-tree model.

- Deligne and Mostow introduced in 1986 most of all presently known commensurability classes of non-arithmetic ball quotients (i.e., quotients of complex hyperbolic n-space by non-arithmetic lattices of $PU(1,n)$). These commensurability classes were mainly distinguished with the aid of two invariants: the associated trace field and the non-compactness of the corresponding ball quotient. Nevertheless, these two invariants are not sufficient to decide whether the precise number (7, 8 or 9?) among the 15 concrete examples of commensurability classes arising from cyclic coverings of the projective line. In 2012, Kappes-Möller [KM12] showed that the techniques related to Kontsevich's formula could be adapted to the setting of ball quotients to provide a *new* invariant (of "Lyapunov exponent" nature) allowing to distinguish the 9 commensurability classes of the 15 concrete examples of ball quotients coming from cyclic coverings.

Partly motivated by this success of Kontsevich's formula for Lyapunov exponents of certain variations of Hodge structures of weight 1 in helping to solve relevant mathematical questions, Möller asked Kontsevich in August 2012 if his formula could be extended to certain variations of Hodge structures of *higher* weights.

After performing numerical experiments with several examples including 14 examples of families of Calabi-Yau 3-folds whose moduli spaces are isomorphic to $\mathbb{C}-\{0,1\}$, Kontsevich discovered that the analog of his formula is *not* always true in the higher weights setting. In particular, he announced in some talks that his formula works for the case of mirror quintics and, actually, his formula is valid exactly for 7 cases among the 14 examples of families of Calabi-Yau 3-folds mentioned above.

In this appendix, we discuss the talks delivered by Kontsevich at IHES in January 2013 and Université Paris-Sud, Orsay in May 2013 by following some notes taken by the author [Kon13]. More concretely, we introduce some basic definitions and notations in Section D.1 including concrete examples of variations of Hodge structures of weight k and the notion of Lyapunov exponents. After that, we quickly review Kontsevich's formula in the context of variations of Hodge structures of weight 1 in Section D.3 below. Then, we reproduce in Section D.4 some of Kontsevich's claims about the valid of his formula for certain variations of Hodge structures of weight 3 with special emphasis in the case of mirror quintic Calabi-Yau 3-folds. Finally, we sketch the proof of a simplicity result (obtained by Eskin and the author) of the Lyapunov exponents associated to mirror quintics (along the lines of Section 4 of [EM12]).

D.1 Examples of variations of Hodge structures of weight k

Let $(X_c)_{c\in C}$ be a one-parameter deformation of compact Kähler manifolds, i.e., $(X_c)_{c\in C}$ is a family of mutually diffeomorphic compact Kähler manifolds whose members X_c are parametrized by $c \in C := \overline{C} - S$, where $\overline{C}$ is a projective complex curve and $S \subset \overline{C}$ is a finite set of points.

Denote by $\mathscr{H}^k$ the local system over C whose fibers are the cohomology groups $H^k_{\mathrm{dR}}(X_c) = H^k(X_c, \mathbb{C})$. The integer lattices $H^k(X_c, \mathbb{Z})$ inside the fibers $H^k(X_c, \mathbb{C}) = \mathbb{C} \otimes H^k(X_c, \mathbb{Z})$ can be naturally identified in a sufficiently small neighborhood of any $c \in C$, and this gives us a flat connection ∇ on $\mathscr{H}^k$ called *Gauss-Manin connection*. In addition to the Gauss-Manin connection, we have the *Hodge decomposition*

$$H^k_{\mathrm{dR}}(X_c) = \bigoplus_{p+q=k} H^{p,q}(X_c)$$

and the *Hodge filtration*

$$\{0\} = F^{k+1} H^k_{\mathrm{dR}}(X_c) \subset \cdots \subset F^p H^k_{\mathrm{dR}}(X_c) \subset \cdots \subset F^0 H^k_{\mathrm{dR}}(X_c) = H^k_{\mathrm{dR}}(X_c)$$

with

$$F^p H^k_{\mathrm{dR}}(X_c) := \bigoplus_{r \geq p} H^{r,k-r}(X_c),$$

where $H^{p,q}(X_c)$ is the space of cohomology classes of type (p,q). The local system $\mathscr{H}^k$ over C together with the Gauss-Manin connection ∇ and the Hodge decomposition (or, equivalently, the Hodge filtration) is a prototypical example of the abstract notion of *variation* of (integral) *Hodge structures* (VHS for short) of weight k.

Example D.1. The third cohomology groups $H^3_{\mathrm{dR}}(X_z)$ of the mirror quintic Calabi-Yau threefolds X_z, $z \in C := \mathbb{C} - \{0,1\}$, described in §3 give rise to a VHS of weight 3.

Example D.2. The natural $SL(2,\mathbb{R})$-action ("*Teichmüller dynamics*") on the moduli space of Abelian differentials on Riemann surfaces of genus $g \geq 1$ leads to several concrete examples of variations of Hodge structures of weight 1 (cf. [EKZ14]).

More precisely, $SL(2,\mathbb{R})$ acts on pairs (M,ω) of a Riemann surface M of genus $g \geq 1$ and an Abelian differential ω on M. If (M,ω) is a *Veech surface*[1], the natural projection C of the $SL(2,\mathbb{R})$-orbit of (M,ω) to the moduli space of curves of genus g (under the map forgetting the Abelian differential) is isomorphic to a *finite area* hyperbolic surface $C \simeq \mathbb{H}/\Gamma$ (that is, Γ is a lattice of $SL(2,\mathbb{R})$ acting on the upper half-plane $\mathbb{H}$ equipped with Poincaré's metric) called *Teichmüller curve*.

The local system $\mathscr{H}^1$ over C whose fibers are the first cohomology groups of the Riemann surfaces of genus $g \geq 1$ parametrized by C gives rise to a VHS of weight 1. (Sometimes $\mathscr{H}^1$ is called *Hodge bundle* in the literature.)

Remark D.1. The recent results of Eskin, Mirzakhani and Mohammadi (see [EM13] and [EMM13]) and Filip (see [Fi13]) allow to generalize the previous example to get variations of Hodge structures of weight 1 parametrized by quasi-projective varieties of any prescribed dimension (obtained from the closures of $SL(2,\mathbb{R})$-orbits in moduli spaces of Abelian differentials).

---

[1] E.g., this is the case when $(M,\omega) \to (\mathbb{C}/(\mathbb{Z} \oplus i\mathbb{Z}), dz)$ is a finite cover ramified only at the origin $0 \in \mathbb{C}/(\mathbb{Z} \oplus i\mathbb{Z})$.

D.2 Lyapunov exponents

From now on, we assume that C is a *hyperbolic* Riemann surface of finite area. This means C is uniformized by $C = \mathbb{H}/\Gamma$, where Γ is a lattice of $SL(2,\mathbb{R})$.

Remark D.2. In this appendix, the upper half-plane $\mathbb{H}$ is *always* endowed with Poincaré metric.

Example D.3. $C = \mathbb{C} - \{0,1\} = \mathbb{C}P^1 - \{0,1,\infty\}$ is a hyperbolic Riemann surface of finite area. Indeed, the Weierstrass function can be used to build an elliptic modular form $\mathbb{H} \to \mathbb{C} - \{0,1\}$ showing that $C = \mathbb{H}/\Gamma_0(2)$, where $\Gamma_0(2)$ is a level 2 congruence subgroup of $SL(2,\mathbb{Z})$.

Example D.4. In general, C is a hyperbolic Riemann surface of finite area whenever $C = \overline{C} - S$, where $\overline{C}$ is a projective complex curve of genus $h \geq 0$ and $S \subset \overline{C}$ is a finite subset of points such that $2h - 2 + \#S > 0$.

Let $\mathscr{E}$ be a vector bundle over a hyperbolic Riemann surface C of finite area and suppose that $\mathscr{E}$ is equipped with a flat connection ∇. For the sake of concreteness, the reader might think of $(\mathscr{E},\nabla)$ as a local system $\mathscr{E} = \mathscr{H}^k$ over C associated to a VHS of weight k together with its Gauss-Manin connection ∇.

In this setting, we have a monodromy representation $\rho : \pi_1(C,c_0) \to GL(N,\mathbb{C})$, where $c_0 \in C$ and $N = \text{rank}(\mathscr{E})$ is the dimension of the fibers of $\mathscr{E}$. This monodromy representation can be used to define a *linear cocycle* (in the sense of Dynamical Systems) over the geodesic flow on (the unit cotangent bundle of) $C = \mathbb{H}/\Gamma$ (with respect to the hyperbolic metric) as follows. We take a very long piece $\{\gamma_v(t)\}_{0\leq t\leq L}$ of trajectory of the geodesic flow starting at a typical cotangent vector v (whose footpoint is nearby $c_0 \in C$). By Poincaré's recurrence theorem, $\gamma_v(L)$ visits any small prescribed neighborhood V of v for adequate arbitrarily large choices of $L \in \mathbb{R}$. By "closing up" the long piece of trajectory $\{\gamma_v(t)\}_{0\leq t\leq L}$ with a small path inside V from $\gamma_v(L)$ to v, we obtain a closed path $\delta(v,L)$ that we think as $\delta(v,L) \in \pi_1(C,c_0)$, and a matrix $A(v,L) = \rho(\delta(v,L)) \in GL(N,\mathbb{C})$. The nomenclature "linear cocycle" for the matrices $A(v,L)$ over the trajectories of the geodesic flow on C comes from the fact that $A(v,L'+L) = A(\gamma_v(L),L') \cdot A(v,L)$.

The dynamics of the linear cocycle (over the geodesic flow on C) associated to $(\mathscr{E},\nabla)$ is described by the celebrated *Oseledets theorem* (also called "multiplicative ergodic theorem" [Os68]). More precisely, this result says that there are real numbers $\lambda_1 > \cdots > \lambda_n$ called *Lyapunov exponents* and, for almost every choice (with respect to the Haar probability measure of $SL(2,\mathbb{R})/\Gamma$) of unit cotangent vector v to $C = \mathbb{H}/\Gamma$, there is a filtration

$$\{0\} = \mathscr{E}_v^{\leq\lambda_{n+1}} \subset \cdots \subset \mathscr{E}_v^{\leq\lambda_l} \subset \cdots \subset \mathscr{E}_v^{\leq\lambda_1} = \mathscr{E}_v$$

of the fiber $\mathscr{E}_v$ of $\mathscr{E}$ at (the footpoint of) v such that

$$\lim_{L\to+\infty} \frac{1}{L} \log \|A(v,L)u_l\|_v = \lambda_l$$

for all $u_l \in \mathscr{E}_v^{\leq\lambda_l} - \mathscr{E}_v^{\leq\lambda_{l+1}}$. Here, $\|.\|_v$ is a "reasonable" choice of norm on $\mathscr{E}_v$ depending measurably on v. (For instance, any choice of $\|.\|_v$ such that $\|.\|_v$ stays "bounded" (uniformly comparable to a fixed norm) near the cusps $s \in S$ of $C = \overline{C} - S$ is "reasonable".)

In other words, Oseledets theorem says that, for almost every v, the singular values of $A(v,L)$ have the form $\exp((\lambda_l + o(1))L)$, $l = 1,\dots,n$, for all L sufficiently large.

Remark D.3. For each $1 \leq l \leq n$, the filtration $\mathscr{E}_v^{\leq\lambda_l}$ depends *measurably* on v in general. In other words, $\{\mathscr{E}_v^{\lambda_l}\}_{1\leq l\leq n}$ is a sort of "fractal Hodge filtration" of the local system $\mathscr{E}$, and, in general, the Lyapunov exponents λ_l have transcendental nature.

Remark D.4. It turns out that the dimension of $\mathscr{E}_v^{\leq\lambda_l}/\mathscr{E}_v^{\leq\lambda_{l+1}}$ does *not* depend on v. The quantity $\dim(\mathscr{E}_v^{\leq\lambda_l}/\mathscr{E}_v^{\leq\lambda_{l+1}})$ is called the *multiplicity* of the Lyapunov exponent λ_l.

Remark D.5. Technically speaking, the Oseledets theorem requires an *L^1-integrability condition* on the linear cocycle. In our setting, this amounts to impose that the spectra of all monodromy matrices $\rho(\gamma)$ associated to small loops γ around the (finitely many) cusps $s \in S$ of $C = \overline{C} - S$ are contained in the unit circle $S^1 = \{z \in \mathbb{C}^\times : |z| = 1\}$.

For the situations we have in mind of $(\mathscr{E},\nabla)$ coming from a VHS of weight k associated to a one-parameter deformation $(X_c)_{c\in C}$ of compact Kähler manifolds, the previous condition is *always* satisfied: indeed, the local monodromy matrix T near a cusp $s \in S$ of a VHS associated to such $(X_c)_{c\in C}$ is *quasi-unipotent* (i.e., $(T^p - 1)^q = 0$ for some integers p and q), see e.g. Théorème 15.15 in Voisin's book [Voi02].

Remark D.6. In his talks, Kontsevich defined the Lyapunov exponents of $(\mathscr{E},\nabla)$ with respect to the *random walk* (Brownian motion) on C. Nevertheless, since the typical random walk tracks a geodesic flow trajectory up to sublinear error (see, e.g., the proof of Theorem 1 in [EM12]), we get the *same* Lyapunov exponents for $(\mathscr{E},\nabla)$ regardless of taking them with respect to the random walk or the geodesic flow.

Remark D.7. From the point of view of numerical experiments, the first digits of the Lyapunov exponents can be accurately computed by the following algorithm. For the sake of concreteness, suppose that C is a finite cover of the modular curve $\mathbb{H}/SL(2,\mathbb{Z})$. In this case, the geodesic flow on C is coded by a finite extension of the *continued fraction algorithm* (because typical geodesics on the modular curve are encoded by the continued fractions). Therefore, we get good approximations for the values of Lyapunov exponents by computing the singular values of the product of monodromy matrices associated to the continued fraction of a random real number.

D.3 Kontsevich's formula in the classical setting

Let $(X_c)_{c\in C}$ be a one-parameter deformation of compact Riemann surfaces of genus $g \geq 1$ (such as those in Example D.2), where C is a hyperbolic Riemann surface of finite area. Denote by $(\mathscr{H}^1, \nabla)$ the pair (Hodge bundle, Gauss-Manin connection) of the corresponding VHS of weight 1.

By definition, $\mathscr{H}^1$ has rank $2g = \dim H^1_{\mathrm{dR}}(X_c)$, so that the monodromy representation associated $(\mathscr{H}^1, \nabla)$ is $\rho : \pi_1(C, c_0) \to GL(2g, \mathbb{C})$. Furthermore, the usual intersection form $\langle \alpha, \beta \rangle = \frac{i}{2} \int \alpha \wedge \overline{\beta}$ on cohomology $H^1_{\mathrm{dR}}(X_c)$ is invariant under the monodromy representation ρ (that is, our VHS $\mathscr{H}^1$ is polarized). Since the intersection form is positive-definite on $H^{1,0}(X_c)$ and negative-definite on $H^{0,1}(X_c)$, the intersection form has signature (g, g) and, hence,

$$\rho : \pi_1(C, c_0) \to U_{\mathbb{C}}(g, g).$$

From this restriction on the monodromy representation ρ, it is possible to show that the Lyapunov exponents of $(\mathscr{H}^1, \nabla)$ have the form

$$\lambda_1 \geq \dots \geq \lambda_g \geq -\lambda_g \geq \dots \geq -\lambda_1$$

where we wrote repeated each λ_l according to its multiplicity. Equivalently, the Lyapunov exponents of $(\mathscr{H}^1, \nabla)$ are symmetric with respect to 0 in the sense that $-\lambda$ is a Lyapunov exponent whenever λ is a Lyapunov exponent.

The transcendental nature of Lyapunov exponents makes that they are very hard to determine in general. Nevertheless, in our current setting[2], Kontsevich discovered (and Forni [Fo01] proved) the following formula for the sum of the non-negative Lyapunov exponents of $(\mathscr{H}^1, \nabla)$:

Theorem D.1 (Kontsevich). *The sum of non-negative Lyapunov exponents is*

$$\lambda_1 + \dots + \lambda_g = 2 \int_C c_1(F^1 \mathscr{H}^1)$$

where $c_1(F^1 \mathscr{H}^1)$ is the first Chern class of (Deligne's extension of) the middle part $F^1 \mathscr{H}^1 = \mathscr{H}^{1,0}$ of the Hodge filtration of $\mathscr{H}^1$.

A detailed proof of this formula can be found at Section 3 of Eskin-Kontsevich-Zorich paper [EKZ14]. For the sake of convenience of the reader, let us just sketch the main observation of Kontsevich lying at the heart of the proof of Theorem D.1 (using a slightly different setting from [EKZ14]).

From the definitions, one can check that the sum $\lambda_1 + \dots + \lambda_g$ of non-negative Lyapunov exponents can be computed from the growth of the norms of polyvectors $\wedge^g A(v, L) \cdot \wedge^g E = \wedge^g E(v, L)$, where $A(v, L) = \rho(\delta(v, L))$, $L \in \mathbb{R}$ large, is a monodromy matrix, and E is a typical g-dimensional subspace inside the isotropic (light)

[2] Actually, Kontsevich's formula holds in a slightly more general context than the one discussed here. See [EKZ14] for more details.

cone of the intersection form $\langle .,. \rangle$ (of signature (g,g)). In order to compute this, Kontsevich notices that, after a simple and clever calculation, one has

$$\log \| \wedge^g E(v,L)\|^2 = \log|\langle \wedge^g E(v,L), \Omega^{g,0}\rangle| + \log|\langle \wedge^g E(v,L), \Omega^{0,g}\rangle| \\ - \log|\langle \Omega^{g,0} \wedge \Omega^{0,g}\rangle|, \qquad \text{(D.1)}$$

where the norm of the polyvector $\wedge^g E(v,L)$ is measured in terms of the *Hodge norm* (of the projection to $\wedge^g H^{1,0}$), $\Omega^{g,0} \in \wedge^g H^{1,0}$, $\Omega^{0,g} \in \wedge^g H^{0,1}$ and the angle between polyvectors is measured using the intersection form.

The main point of this remarkable observation of Kontsevich is the fact that the first two terms $\log|\langle \wedge^g E(v,L), \Omega^{g,0}\rangle|$ and $\log|\langle \wedge^g E(v,L), \Omega^{0,g}\rangle|$ of the right-hand side of (D.1) are *harmonic* functions of the footpoint of v in the hyperbolic Riemann surface C (because $\langle \wedge^g E(v,L), \Omega^{g,0}\rangle$ is holomorphic and $\langle \wedge^g E(v,L), \Omega^{0,g}\rangle$ is antiholomorphic), while the last term $\log|\langle \Omega^{g,0} \wedge \Omega^{0,g}\rangle|$ (the "curvature" of $F^1H^1 = H^{1,0}$) is *independent* of E.

By exploiting this, one can show from harmonicity that the first two terms of the right-hand side of (D.1) do not contribute to the computation of the sum of non-negative Lyapunov exponents, so that $\lambda_1 + \cdots + \lambda_g$ can be expressed only in terms of $\log|\langle \Omega^{g,0} \wedge \Omega^{0,g}\rangle|$, i.e., the first Chern class ("curvature") of $F^1\mathscr{H}^1$.

Remark D.8. The formula in Theorem D.1 differs slightly from the equation in Subsection 3.3 of [EKZ14]. In fact, the precise formula depends on the normalization of the curvature of the hyperbolic metric on C, and, for the sake of simplicity, our normalization is different from [EKZ14].

Remark D.9. A striking consequence (from the point of view of Dynamical Systems) of Kontsevich's formula is the *rationality* of the sum of non-negative Lyapunov exponents: indeed, $\lambda_1 + \cdots + \lambda_g = 2\int_C c_1(F^1\mathscr{H}^1) = 2\deg(F^1\mathscr{H}^1) \in \mathbb{Q}$, where $\deg(F^1\mathscr{H}^1)$ is the orbifold degree of (Deligne's extension of) the holomorphic subbundle $F^1\mathscr{H}^1$ over C. (This is in sharp contrast with the transcendental nature of Lyapunov exponents for *general* dynamical systems.)

Remark D.10. Kontsevich's formula has some interesting combinatorial applications along the lines of appendices A and B of [EKZ14] (and, in fact, Kontsevich does not know how to get these combinatorial facts without using his formula).

For example, for each N, consider the action of the free group F_2 on the space $S_N \times S_N$ of pairs of permutations $\{1,\dots,N\}$ induced by Nielsen transformations $(a,b) \mapsto (a,ab)$ and $(a,b) \mapsto (ab,b)$. Then, for *all* $N \geq 9$ and for *all* orbit $\mathscr{O}$ of F_2 acting on $S_N \times S_N$ containing a pair of permutations $(a,b) \in \mathscr{O}$ such that a and b act transitively on $\{1,\dots,N\}$ and the cycle structure of the commutator $[a,b] = aba^{-1}b^{-1}$ is $(1^{N-9}, 9^1)$ (i.e., $[a,b]$ has a single non-trivial cycle of size 9), the quantity

$$\frac{1}{\#\mathscr{O}} \sum_{(c,d)\in\mathscr{O}} \sum_{\substack{\sigma \text{ is} \\ \text{cycle of } c}} \frac{1}{\text{length of } \sigma}$$

takes one of the following *three* values:

$$\frac{55}{27}, \quad \frac{40}{27} \quad \text{or} \quad \frac{37}{27}.$$

Moreover, *each* of these three values is realized by some orbit $\mathcal{O}$ as above.

D.4 Kontsevich's formula in Calabi-Yau 3-folds setting

Möller asked (in August 2012) Kontsevich whether his formula in Theorem D.1 could be generalized to VHS of *higher* weights.

In this direction, Kontsevich decided to test the 14 conjugacy classes of (real) VHS of weight 3 associated to families of Calabi-Yau threefolds with moduli space isomorphic to $C := \mathbb{C} - \{0,1\} = \mathbb{C}P^1 - \{0,1,\infty\}$.

More precisely, consider[3] a one-parameter deformation $(X_c)_{c\in C}$ of Calabi-Yau threefolds with Hodge numbers $h^{p,q} = \dim H^{p,q}(X_c) = 1$ for $p+q=3$ such that their third cohomology groups lead to (integral) VHS $(\mathscr{H}^3, \nabla)$ of weight 3 with irreducible monodromy representation

$$\rho : \pi_1(\mathbb{C}P^1 - \{0,1,\infty\}, i) \to Sp(4,\mathbb{Z})$$

which is maximal unipotent at $z=0$ (MUM singularity), unipotent of rank 1 at $z=1$ (conifold singularity), and quasi-unipotent at $z=\infty$.

In other terms, denoting by γ_0, γ_1 and γ_∞ small loops around 0, 1 and ∞, we require that $M_0 = \rho(\gamma_0) = Id + N_0$, where N_0 has rank 3, $M_1 = Id + N_1$, where N_1 has rank 1, and $M_\infty = \rho(\gamma_\infty) = (M_0 M_1)^{-1}$ satisfy $(M_\infty^k - Id)^l = 0$ for some integers k and l.

It is known (cf. [DM06]) that the conjugacy class of ρ (as a *real* representation) falls into 14 categories: more precisely, the characteristic polynomial of M_∞^{-1} has the form

$$x^4 + (a-4)x^3 + (6-2a+m)x^2 + (a-4)x + 1,$$

the integers m, a are a complete invariant of the (real) conjugacy class of ρ, and the pair (m,a) take one of the following 14 values:

$$(1,2),(1,3),(1,4),(2,3),(2,4),(3,4),(4,4),$$

$$(4,5),(5,5),(6,5),(8,6),(9,6),(12,7),(16,8).$$

In §7.2 the notation $(d,k) = (m,a)$ has been used.

Remark D.11. The family of mirror quintics described in §3 corresponds to the parameters $(m,a) = (5,5)$: this follows from the fact that $x^4 + x^3 + x^2 + x + 1$ is the characteristic polynomial of the product $M_0 \cdot M_1$ of the monodromy matrices M_0 and M_1 calculated in §2.7.

[3] Such one-parameter deformations of Calabi-Yau threefolds come from *mirror symmetry*, or more precisely, from Batyrev-Borisov mirrors of Calabi-Yau complete intersections in weighted projective spaces. See., e.g., [DM06] for more details.

The Lyapunov exponents of each of these $(\mathscr{H}^3, \nabla)$ have the form

$$\lambda_1 \geq \lambda_2 \geq -\lambda_2 \geq -\lambda_1.$$

Indeed, this happens because $\mathscr{H}^3$ has rank $4 = \sum\limits_{p+q=3} h^{p,q}$ and the intersection form $\langle .,. \rangle$ (introduced in §2) has signature $(2,2)$.

After running numerical experiments, Kontsevich saw that the analog of his formula (in Theorem D.1) for the sum $\lambda_1 + \lambda_2$ of non-negative exponents $(\mathscr{H}^3, \nabla)$ is true *exactly* in 7 of the 14 cases above:

Claim (Kontsevich). In the setting above, denote the spectrum of $M_0 \cdot M_1 = M_\infty^{-1}$ by

$$\{\exp(2\pi i \alpha_1), \exp(2\pi i \alpha_2), \exp(-2\pi i \alpha_1), \exp(-2\pi i \alpha_2)\}$$

where $0 < \alpha_1 \leq \alpha_2 \leq 1/2$. Then, one of the following two possibilities happens:

(a) "**Good cases**": if $3\alpha_2 \geq \alpha_1 + 1$, we have $\lambda_1 + \lambda_2 = 2(\alpha_1 + \alpha_2) = 2\int_C c_1(F^2\mathscr{H}^3)$, i.e., the sum of non-negative Lyapunov exponents is determined by the first Chern class of the middle part $F^2\mathscr{H}^3 = \mathscr{H}^{3,0} \oplus \mathscr{H}^{2,1}$ of the Hodge filtration of $\mathscr{H}^3$;
(b) "**Bad cases**": otherwise, $\lambda_1 + \lambda_2 > 2(\alpha_1 + \alpha_2) = 2\int_C c_1(F^2\mathscr{H}^3)$.

The values of (α_1, α_2) for each of the 14 values of (m,a) are listed in §7.2 (note that $(d,k) = (m,a)$). A direct inspection of this table shows that the inequality $3\alpha_2 \geq \alpha_1 + 1$, i.e., Kontsevich's formula, is satisfied for exactly 7 of the 14 possible choices of (m,a). Furthermore, in the case $(m,a) = (5,5)$ of mirror quintics, an inspection of this table (or, alternatively, a direct application of Remark D.11) reveals that $\alpha_1 = 1/5$ and $\alpha_2 = 2/5$, so that $3\alpha_2 = \alpha_1 + 1$. In other words, Kontsevich's formula applies in the context of mirror quintics and it gives that the sum of non-negative Lyapunov exponents is $\lambda_1 + \lambda_2 = 6/5$.

Remark D.12. Interestingly enough, the good cases for Kontsevich's formula are precisely those where the image of the monodromy representation ρ is a *thin subgroup* of $Sp(4,\mathbb{Z})$ in Sarnak's terminology in [Sar12] [4], while the bad cases for Kontsevich's formula correspond exactly to the image of ρ being an arithmetic (i.e., finite-index) subgroups of $Sp(4,\mathbb{Z})$.

In his talk, Kontsevich said that he does not think this is just a coincidence (and he suspects that sums of Lyapunov exponents might be helpful in studying thin groups), but he has no results in this direction.

Besides the strong numerical evidence for Claim above, Kontsevich offered the following *heuristic* explanation for his claim.

We start the argument in the same way as in the sketch of proof of Theorem D.1: the sum $\lambda_1 + \lambda_2$ is given by the growth under the second exterior power of the monodromy representation of polyvectors associated to typical 2-dimensional subspaces E inside the isotropic cone of the intersection form on the third cohomology.

[4] That is, an infinite index subgroup of $Sp(4,\mathbb{Z})$ which is Zariski dense in $Sp(4,\mathbb{R})$.

Next, we consider the logarithm of the square of the norm of such polyvectors, and we try to write it as a sum of an harmonic function and the "curvature" of $F^2\mathscr{H}^3$. By doing so, this time we *might* have a problem with the harmonic function when E is *not* transversal to $F^2\mathscr{H}^3 = \mathscr{H}^{3,0} \oplus \mathscr{H}^{2,1}$. More precisely, the intersection form is positive-definite on $\mathscr{H}^{3,0}$ and negative-definite on $\mathscr{H}^{2,1}$, so that the isotropic subspace E could be non-transversal to $F^2\mathscr{H}^3$. If this non-transversality happens, the harmonic function (obtained as the sum of the logarithms of the moduli of the angles of the polyvector $\wedge^2 E$ with $F^2\mathscr{H}^3$ and $\overline{F^2\mathscr{H}^3}$) develops a logarithm singularity leading to a positive current giving an extra contribution for the formula of $\lambda_1 + \lambda_2$ in terms of the curvature of $F^2\mathscr{H}^3$. In summary, the non-transversality between (the Oseledets unstable subspace) E and $F^2\mathscr{H}^3$ is responsible for

$$\lambda_1 + \lambda_2 > 2\int_C c_1(F^2\mathscr{H}^3).$$

Alternatively, consider the action of the monodromy representation ρ on the Grassmanian of isotropic 2-planes. Suppose that we can construct[5] a monodromy invariant probability measure μ on this Grassmanian such that its support is transversal to $F^2\mathscr{H}^3$, or, equivalently, assume that we can build up a monodromy invariant *non-trivial* open subset U of the Grassmanian containing $F^2\mathscr{H}^3$. In this case, the arguments of the proof of Theorem D.1 apply in our current setting to show that

$$\lambda_1 + \lambda_2 = 2\int_C c_1(F^2\mathscr{H}^3).$$

Remark D.13. In the case of mirror quintics, Kontsevich told that this heuristic argument can be reformulated as follows. By writing down this non-transversality condition, one sees that an open set U as above can be constructed if Problem 2 in §11 has a positive solution, i.e., if the analytic continuations of the Wronskian determinant $\psi_0\theta\psi_1 - \psi_1\theta\psi_0$ in $\mathbb{C}P^1 - \{0,1,\infty\}$ has no zeroes, where ψ_0 and ψ_1 are the holomorphic and logarithmic solutions at $z = 0$ of the differential equation (2.25). In particular, Kontsevich claimed that his several numerical experiments are in favor of a positive solution to Problem 2, so that this gives further evidence for the fact that mirror quintics fall in the "good case" of Claim in §D.4.

Remark D.14. Very recently, Filip [Fi14] studied (with the help of Kuga-Satake construction) the analog of Kontsevich's formula for the *top* Lyapunov exponent λ_1 in the context of VHS of weight 2 coming from families of K3 surfaces.

Remark D.15. In view of the nice applications (of Delecroix-Hubert-Lelièvre [DHL14], Kappes-Möller [KM12], etc.) of Lyapunov exponents for VHS of weight 1, it is a natural question to find relevant applications in Algebraic Geometry (or Dynamical Systems, Number Theory, etc.) of Lyapunov exponents for VHS of higher weights.

---

[5] The reader should think of this measure as Dirac masses on Oseledets unstable subspaces.

D.5 Simplicity of Lyapunov exponents of mirror quintics

The formula of Kontsevich Claim in §D.4 allows to infer that $\lambda_1 + \lambda_2 > 0$, and, hence, $\lambda_1 > 0$, in all 14 cases of VHS of weight 3 considered above. On the other hand, it gives no clue whether $\lambda_1 > \lambda_2$ and/or $\lambda_2 > 0$.

In this direction, Eskin and the author [EM12] showed that:

Theorem D.2. *In the case of mirror quintics,* $\lambda_1 > \lambda_2 > 0$.

The proof of this result goes as follows (see Section 4 of [EM12] for more details). Let $\rho : \pi_1(\mathbb{C}P^1 - \{0,1,\infty\}, i) \to Sp(4,\mathbb{Z})$ be the monodromy representation associated to the one-parameter family of mirror quintic Calabi-Yau threefolds.

By the main result (Theorem 1) of [EM12] and the "Galois-theoretical" simplicity criterion[6] of Theorem 5.4 of [MMY], we have:

Theorem D.3. *Let $\mathcal{G}$ be the monoid generated by the monodromy matrices $M_0 = \rho(\gamma_0)$ and $M_1 = \rho(\gamma_1)$. Suppose that $\mathcal{G}$ contains two matrices $A, B \in Sp(4,\mathbb{Z})$. Denote by $P(x) = x^4 + a(P)x^3 + b(P)x^2 + a(P)x + 1$ and $Q(x) = x^4 + a(Q)x^3 + b(Q)x^2 + a(Q)x + 1$ the characteristic polynomials of A and B. Suppose that the discriminants*

- $\Delta_1(P) := a(P)^2 - 4(b(P)-2)$, $\Delta_2(P) := (b(P)+2)^2 - 4a(P)^2$, $\Delta_3(P) := \Delta_1(P) \cdot \Delta_2(P)$,
- $\Delta_1(Q) := a(Q)^2 - 4(b(Q)-2)$, $\Delta_2(Q) := (b(Q)+2)^2 - 4a(Q)^2$, $\Delta_3(Q) := \Delta_1(Q) \cdot \Delta_2(Q)$, *and*
- $\Delta_i(P) \cdot \Delta_j(Q)$, $1 \le i, j \le 3$

are positive integers that are not squares. Then, $\lambda_1 > \lambda_2 > 0$.

By Theorem D.3, the proof of Theorem D.2 is reduced to understand the monoid $\mathcal{G}$. For this sake, let us recall that the monodromy matrices M_0 and M_1 were computed (in an appropriate basis) in (2.26):

$$M_0 = \begin{pmatrix} 1 & 1 & 0 & 0 \\ 0 & 1 & 0 & 0 \\ 5 & 5 & 1 & 0 \\ 0 & -5 & -1 & 1 \end{pmatrix}, \quad M_1 = \begin{pmatrix} 1 & 0 & 0 & 0 \\ 0 & 1 & 0 & 1 \\ 0 & 0 & 1 & 0 \\ 0 & 0 & 0 & 1 \end{pmatrix}.$$

We consider the matrices $A := M_0^3 \cdot M_1 \in \mathcal{G}$ and $B := M_0^4 \cdot M_1 \in \mathcal{G}$. A direct computation shows that their respective characteristic polynomials are

$$P(x) = x^4 + 31x^3 + 71x^2 + 31x + 1$$

and

$$Q(x) = x^4 + 66x^3 + 186x^2 + 66x + 1.$$

---

[6] The conditions of Theorem D.3 ensure that the characteristic polynomials of A and B are irreducible, their roots are real, their Galois groups are hyperoctahedral (of order 8), and their splitting fields are disjoint. This justifies the nomenclature "Galois-theoretical" simplicity criterion.

The corresponding discriminants are

$$\Delta_1(P) = 685 = 5 \times 137, \quad \Delta_2(P) = 1485 = 3^3 \times 5 \times 11$$

and

$$\Delta_1(Q) = 3620 = 2^2 \times 5 \times 181, \quad \Delta_2(P) = 17920 = 2^9 \times 5 \times 7.$$

It follows that the monoid $\mathscr{G}$ generated by M_0 and M_1 satisfies the assumptions of Theorem D.3, so that the proof of Theorem D.2 is complete.

Remark D.16. Even though the author has not verified all details, he thinks that the simplicity statement of Theorem D.2 can be extended to *all* 14 cases of VHS of weight 3 in Section D.4. In fact, this would be a consequence of the main result (Theorem 1) of [EM12] together with the Zariski density of the image of the monodromy representation ρ in $Sp(4,\mathbb{R})$.

References

ABKv2. Mina Aganagic, Vincent Bouchard, and Albrecht Klemm. Topological strings and (almost) modular forms. *Comm. Math. Phys.*, 277(3):771–819, 2008, arXiv:hep-th/0607100v2.

AL07. Murad Alim and Jean Dominique Länge. Polynomial structure of the (open) topological string partition function. *J. High Energy Phys.*, (10):045, 13, 2007.

Ali13. Murad Alim. Lectures on Mirror Symmetry and Topological String Theory. *Open Problems and Surveys of Contemporary Mathematics*, 6:1, 2013.

Ali14. Murad Alim. Lie algebra and differential rings of $tt*$ geometries of Calabi-Yau sigma models. *Preprint*, 2014.

AMSY14. Murad Alim, Hossein Movasati, Emanuel Scheidegger, and Shing-Tung Yau. Gauss-Manin connection in disguise: Calabi-Yau threefolds. *Comm. Math. Phys.*, 344(3):889–914, 2016.

BCDT01. Christophe Breuil, Brian Conrad, Fred Diamond, and Richard Taylor. On the modularity of elliptic curves over $\mathbf{Q}$: wild 3-adic exercises. *J. Amer. Math. Soc.*, 14(4):843–939 (electronic), 2001.

BCOV93. M. Bershadsky, S. Cecotti, H. Ooguri, and C. Vafa. Holomorphic anomalies in topological field theories. *Nuclear Phys. B*, 405(2-3):279–304, 1993.

BCOV94. M. Bershadsky, S. Cecotti, H. Ooguri, and C. Vafa. Kodaira-Spencer theory of gravity and exact results for quantum string amplitudes. *Comm. Math. Phys.*, 165(2):311–427, 1994.

BH89. F. Beukers and G. Heckman. Monodromy for the hypergeometric function ${}_nF_{n-1}$. *Invent. Math.*, 95(2):325–354, 1989.

Bor92. Richard E. Borcherds. Monstrous moonshine and monstrous Lie superalgebras. *Invent. Math.*, 109(2):405–444, 1992.

BT14. Christopher Brav and Hugh Thomas. Thin monodromy in Sp(4). *Compos. Math.*, 150(3):333–343, 2014.

Bui00. Alexandru Buium. Differential modular forms. *J. Reine Angew. Math.*, 520:95–167, 2000.

CDF$^+$97. A. Ceresole, R. D'Auria, S. Ferrara, W. Lerche, J. Louis, and T. Regge. Picard-Fuchs equations, special geometry and target space duality. In *Mirror symmetry, II*, volume 1 of *AMS/IP Stud. Adv. Math.*, pages 281–353. Amer. Math. Soc., Providence, RI, 1997.

CdlO91. Philip Candelas and Xenia de la Ossa. Moduli Space of Calabi-Yau Manifolds. *Nucl.Phys.*, B355:455–481, 1991.

CdlOGP91. Philip Candelas, Xenia C. de la Ossa, Paul S. Green, and Linda Parkes. A pair of Calabi-Yau manifolds as an exactly soluble superconformal theory. *Nuclear Phys. B*, 359(1):21–74, 1991.

CdlORV03. Philip Candelas, Xenia de la Ossa, and Fernando Rodriguez-Villegas. Calabi-Yau manifolds over finite fields. II. In *Calabi-Yau varieties and mirror symmetry (Toronto, ON, 2001)*, volume 38 of *Fields Inst. Commun.*, pages 121–157. Amer. Math. Soc., Providence, RI, 2003.

C13. D. Chen. Strata of abelian differentials and the Teichmüller dynamics. *J. Mod. Dyn.*, 7:135–152, 2013.

CM12. D. Chen and M. Möller. Nonvarying sums of Lyapunov exponents of Abelian differentials in low genus. *Geom. Topol.*, 16:2427–2479, 2012.

CK99. David A. Cox and Sheldon Katz. *Mirror symmetry and algebraic geometry*, volume 68 of *Mathematical Surveys and Monographs*. American Mathematical Society, Providence, RI, 1999.

Cre97. J. E. Cremona. *Algorithms for modular elliptic curves.* Cambridge University Press, Cambridge, 1997.

CvS13. Slawomir Cynk and Duco van Straten. Calabi-Yau conifold expansions. In *Arithmetic and Geometry of K3 Surfaces and CalabiCYau Threefolds*, pages 499–515. Springer, New York, 2013.

CYY08. Yao-Han Chen, Yifan Yang, and Noriko Yui. Monodromy of Picard-Fuchs differential equations for Calabi-Yau threefolds. *J. Reine Angew. Math.*, 616:167–203, 2008. With an appendix by Cord Erdenberger.

DHL14. V. Delecroix, P. Hubert, and S. Lelièvre. Diffusion for the periodic wind-tree model. *Ann. Sci. Éc. Norm. Supér. (4)*, 47(6):1058–1110, 2014.

Del97. P. Deligne. Local behavior of Hodge structures at infinity. In *Mirror symmetry, II*, volume 1 of *AMS/IP Stud. Adv. Math.*, pages 683–699. Amer. Math. Soc., Providence, RI, 1997.

Del09. Pierre Deligne. Private letter, [Supp Item 59], 2009.

DGMS13. C. Doran, T. Gannon, H. Movasati, and K. Shokri. Automorphic forms for triangle groups. *Communications in Number Theory and Physics*, 7(4):689–737, 2013.

DGS94. Bernard Dwork, Giovanni Gerotto, and Francis J. Sullivan. *An introduction to G-functions*, volume 133 of *Annals of Mathematics Studies*. Princeton University Press, Princeton, NJ, 1994.

Dij95. Robbert Dijkgraaf. Mirror symmetry and elliptic curves. In *The moduli space of curves (Texel Island, 1994)*, volume 129 of *Progr. Math.*, pages 149–163. Birkhäuser Boston, Boston, MA, 1995.

Dij97. Robbert Dijkgraaf. Chiral deformations of conformal field theories. *Nuclear Phys. B*, 493(3):588–612, 1997.

DM06. Charles F. Doran and John W. Morgan. Mirror symmetry and integral variations of Hodge structure underlying one-parameter families of Calabi-Yau threefolds. In *Mirror symmetry. V*, volume 38 of *AMS/IP Stud. Adv. Math.*, pages 517–537. Amer. Math. Soc., Providence, RI, 2006.

DMOS82. Pierre Deligne, James S. Milne, Arthur Ogus, and Kuang-yen Shih. *Hodge cycles, motives, and Shimura varieties*, volume 900 of *Lecture Notes in Mathematics*. Springer-Verlag, Berlin, 1982. Philosophical Studies Series in Philosophy, 20.

DS05. Fred Diamond and Jerry Shurman. *A first course in modular forms*, volume 228 of *Graduate Texts in Mathematics*. Springer-Verlag, New York, 2005.

Dwo69. B. Dwork. p-adic cycles. *Inst. Hautes Études Sci. Publ. Math.*, (37):27–115, 1969.

Dwo73. B. Dwork. On p-adic differential equations. IV. Generalized hypergeometric functions as p-adic analytic functions in one variable. *Ann. Sci. École Norm. Sup. (4)*, 6:295–315, 1973.

EKZ14. A. Eskin, M. Kontsevich, and A. Zorich. Sum of Lyapunov exponents of the Hodge bundle with respect to the Teichmüller geodesic flow. *Publ. Math. Inst. Hautes Études Sci.*, 120:207–333, 2014.

EM12. A. Eskin and C. Matheus. A coding-free simplicity criterion for the Lyapunov exponents of Teichmüller curves. *Geometriae Dedicata*, 179(1):45-67, 2015.

EM13. A. Eskin and M. Mirzakhani. Invariant and stationary measures for the $SL(2,\mathbb{R})$ action on moduli space. arXiv:1302.3320, 2013.

EMM13. A. Eskin, M. Mirzakhani, and A. Mohammadi. Isolation, equidistribution, and orbit closures for the $SL(2,\mathbb{R})$ action on moduli space. arXiv:1305.3015, 2013.

Fi13. S. Filip. Splitting mixed Hodge structures over affine invariant manifolds. arXiv:1311.2350, 2013.

Fi14. S. Filip. Families of K3 surfaces and Lyapunov exponents. arXiv:1412.1779, 2014.

Fo01. G. Forni. Deviation of ergodic averages for area-preserving flows on surfaces of higher genus. *Ann. of Math. (2)*, 155:1–103, 2002.

Fre83. Eberhard Freitag. *Siegelsche Modulfunktionen*, volume 254 of *Grundlehren der Mathematischen Wissenschaften*. Springer-Verlag, Berlin, 1983.

Gan06. Terry Gannon. *Moonshine beyond the Monster*. Cambridge Monographs on Mathematical Physics. Cambridge University Press, Cambridge, 2006. The bridge connecting algebra, modular forms and physics.

GAZ10. Duco van Straten Gert Almkvist, Christian van Enckevort, and Wadim Zudilin. Tables of Calabi-Yau equations. *Preprint, arXiv:math/0507430*, 2010.

GH94. Phillip Griffiths and Joseph Harris. *Principles of algebraic geometry*. Wiley Classics Library. John Wiley & Sons Inc., New York, 1994. Reprint of the 1978 original.

GPS01. G.-M. Greuel, G. Pfister, and H. Schönemann. SINGULAR 2.0. A Computer Algebra System for Polynomial Computations. Centre for Computer Algebra, University of Kaiserslautern, 2001. `http://www.singular.uni-kl.de`.

Gri70. Phillip A. Griffiths. Periods of integrals on algebraic manifolds: Summary of main results and discussion of open problems. *Bull. Amer. Math. Soc.*, 76:228–296, 1970.

Gro66. Alexander Grothendieck. On the de Rham cohomology of algebraic varieties. *Inst. Hautes Études Sci. Publ. Math.*, (29):95–103, 1966.

Gui07. Adolfo Guillot. Sur les équations d'Halphen et les actions de $SL_2(\mathbf{C})$. *Publ. Math. Inst. Hautes Études Sci.*, (105):221–294, 2007.

GVv1. Rajesh Gopakumar and Cumrun Vafa. M-theory and topological strings–II. arXiv:hep-th/9812127v1, 1998.

Hal81. G. H. Halphen. Sur des fonctions qui proviennent de l'équation de gauss. *C. R. Acad. Sci Paris*, 92:856–859, 1881.

Hec36. E. Hecke. Über die Bestimmung Dirichletscher Reihen durch ihre Funktionalgleichung. *Math. Ann.*, 112(1):664–699, 1936.

HKK$^+$03. Kentaro Hori, Sheldon Katz, Albrecht Klemm, Rahul Pandharipande, Richard Thomas, Cumrun Vafa, Ravi Vakil, and Eric Zaslow. *Mirror symmetry*, volume 1 of *Clay Mathematics Monographs*. American Mathematical Society, Providence, RI; Clay Mathematics Institute, Cambridge, MA, 2003. With a preface by Vafa.

HKQ09. M.-x. Huang, A. Klemm, and S. Quackenbush. Topological string theory on compact Calabi-Yau: modularity and boundary conditions. In *Homological mirror symmetry*, volume 757 of *Lecture Notes in Phys.*, pages 45–102. Springer, Berlin, 2009.

HLY96. Shinobu Hosono, Bong H. Lian, and Shing-Tung Yau. Gkz-generalized hypergeometric systems in mirror symmetry of Calabi-Yau hypersurfaces. *Commun. Math. Phys.*, 182:535–578, 1996.

Hos08. Shinobu Hosono. BCOV ring and holomorphic anomaly equation. In *New developments in algebraic geometry, integrable systems and mirror symmetry (RIMS, Kyoto)*, volume 59 of *Adv. Stud. Pure Math.*, pages 79–110. Math. Soc. Japan, Tokyo, 2008.

KM12. A. Kappes and M. Möller. Lyapunov spectrum of ball quotients with applications to commensurability questions. *Duke Mathematical Journal*, 165(1):1-66, 2016.

KKRS05. Albrecht Klemm, Maximilian Kreuzer, Erwin Riegler, and Emanuel Scheidegger. Topological string amplitudes, complete intersection Calabi-Yau spaces and threshold corrections. *J. High Energy Phys.*, (5):023, 116 pp. (electronic), 2005.

KKV99. Sheldon Katz, Albrecht Klemm, and Cumrun Vafa. M-theory, topological strings and spinning black holes. *Adv. Theor. Math. Phys.*, 3(5):1445–1537, 1999.

KO68. Nicholas M. Katz and Tadao Oda. On the differentiation of de Rham cohomology classes with respect to parameters. *J. Math. Kyoto Univ.*, 8:199–213, 1968.

Kob93. Neal Koblitz. *Introduction to elliptic curves and modular forms*, volume 97 of *Graduate Texts in Mathematics*. Springer-Verlag, New York, second edition, 1993.

Kon96. M. Kontsevich. Lyapunov exponents and Hodge theory. In *The mathematical beauty of physics (Saclay, 1996)*, pages 318–332. Adv. Ser. Math. Phys., 24. World Sci. Publ., River Edge, NJ, 1997.

Kon13. M. Kontsevich. Lyapunov exponents for variations of Hodge structures. http://matheuscmss.wordpress.com/2013/01/09/maxim-kontsevichs-talk-on-lyapunov- exponents-for-variations-of-hodge-structures/.

Kon95. Maxim Kontsevich. Homological algebra of mirror symmetry. In *Proceedings of the International Congress of Mathematicians, Vol. 1, 2 (Zürich, 1994)*, pages 120–139. Birkhäuser, Basel, 1995.

KRNT. Christian Krattenthaler and Tanguy Rivoal. On the integrality of the Taylor coefficients of mirror maps. *Duke Math. J.*, 151(2):175–218, 2010, arXiv:0907.2577v2[math.NT].

KSVv3. Maxim Kontsevich, Albert Schwarz, and Vadim Vologodsky. Integrality of instanton numbers and p-adic B-model. *Phys. Lett. B*, 637(1-2):97–101, 2006, arXiv:hep-th/0603106v3.

KU02. Kazuya Kato and Sampei Usui. Borel-Serre spaces and spaces of SL(2)-orbits. In *Algebraic geometry 2000, Azumino (Hotaka)*, volume 36 of *Adv. Stud. Pure Math.*, pages 321–382. Math. Soc. Japan, Tokyo, 2002.

KZ95. Masanobu Kaneko and Don Zagier. A generalized Jacobi theta function and quasimodular forms. In *The moduli space of curves (Texel Island, 1994)*, volume 129 of *Progr. Math.*, pages 165–172. Birkhäuser Boston, Boston, MA, 1995.

KZ01. Maxim Kontsevich and Don Zagier. Periods. In *Mathematics unlimited—2001 and beyond*, pages 771–808. Springer, Berlin, 2001.

Lan95a. Serge Lang. Introduction to modular forms, volume 222 of *Grundlehren der Mathematischen Wissenschaften*. Springer-Verlag, Berlin, 1995. With appendixes by D. Zagier and Walter Feit, Corrected reprint of the 1976 original.

Lan95b. Serge Lang. Some history of the Shimura-Taniyama conjecture. *Notices Amer. Math. Soc.*, 42(11):1301–1307, 1995.

Lew99. James D. Lewis. A survey of the Hodge conjecture, volume 10 of *CRM Monograph Series*. American Mathematical Society, Providence, RI, second edition, 1999. Appendix B by B. Brent Gordon.

LS04. Zhiqin Lu and Xiaofeng Sun. Weil-Petersson geometry on moduli space of polarized Calabi-Yau manifolds. *J. Inst. Math. Jussieu*, 3(2):185–229, 2004.

LY96a. Bong H. Lian and Shing-Tung Yau. Arithmetic properties of mirror map and quantum coupling. *Comm. Math. Phys.*, 176(1):163–191, 1996.

LY96b. Bong H. Lian and Shing-Tung Yau. Mirror maps, modular relations and hypergeometric series I, II. *In Proceedings of the Conference on S-duality in Triste ICTP 1995, Nucl. Phys.*, pages 248–262, 1996.

LY98. Bong H. Lian and Shing-Tung Yau. Integrality of certain exponential series. In *Algebra and geometry (Taipei, 1995), Lect. Alg. Geom.*, pages 215–227. Internat. Press, Cambridge, MA, 1998.

MMY. C. Matheus, M. Möller, and J.-C. Yoccoz. A criterion for the simplicity of the Lyapunov spectrum of square-tiled surfaces. *Inventiones mathematicae*, 202(1):333-425, 2015.

Miy89. Toshitsune Miyake. *Modular forms*. Springer-Verlag, Berlin, 1989. Translated from the Japanese by Yoshitaka Maeda.

MNOP06. D. Maulik, N. Nekrasov, A. Okounkov, and R. Pandharipande. Gromov-Witten theory and Donaldson-Thomas theory. I. *Compos. Math.*, 142(5):1263–1285, 2006.

Mor92. David R. Morrison. Picard-Fuchs equations and mirror maps for hypersurfaces. In *Essays on mirror manifolds*, pages 241–264. Int. Press, Hong Kong, 1992. arXiv:alg-geom/9202026v1.

Mor93. David R. Morrison. Mirror symmetry and rational curves on quintic threefolds: a guide for mathematicians. *J. Amer. Math. Soc.*, 6(1):223–247, 1993.

Mov08a. Hossein Movasati. On differential modular forms and some analytic relations between Eisenstein series. *Ramanujan J.*, 17(1):53–76, 2008. arXiv:math/0610861v1[math.NT].

Mov08b. Hossein Movasati. On elliptic modular foliations. *Indag. Math. (N.S.)*, 19(2):263–286, 2008. arXiv:0806.3926v1[math.AG].

Mov11a. Hossein Movasati. Eisenstein type series for Calabi-Yau varieties. *Nuclear Physics B*, 847:460–484, 2011.

Mov11b. Hossein Movasati. Multiple Integrals and Modular Differential Equations. In *28th Brazilian Mathematics Colloquium*. Instituto de Matemática Pura e Aplicada, IMPA, 2011.

Mov12a. Hossein Movasati. On Ramanujan relations between Eisenstein series. *Manuscripta Mathematicae*, 139(3-4):495–514, 2012.

Mov12b. Hossein Movasati. Quasi modular forms attached to elliptic curves, I. *Annales Mathmatique Blaise Pascal*, 19:307–377, 2012.

Mov13. Hossein Movasati. Quasi modular forms attached to Hodge structures. *Fields Communication Series, 67*, 2013.

Mov15. Hossein Movasati. Modular-type functions attached to mirror quintic Calabi-Yau varieties. *Math. Zeit.*, 281, 3: 907–929, 2015.

MR05. François Martin and Emmanuel Royer. Formes modulaires et périodes. In *Formes modulaires et transcendance*, volume 12 of *Sémin. Congr.*, pages 1–117. Soc. Math. France, Paris, 2005.

Os68. V. I. Osedelets. A multiplicative ergodic theorem. Characteristic Ljapunov, exponents of dynamical systems. *Trudy Moskov. Mat. Obšč*, 19:179–210, 1968.

Pan98. Rahul Pandharipande. Rational curves on hypersurfaces (after A. Givental). *Astérisque*, (252):Exp. No. 848, 5, 307–340, 1998. Séminaire Bourbaki. Vol. 1997/98.

Rib04. Kenneth A. Ribet. Abelian varieties over **Q** and modular forms. In *Modular curves and abelian varieties*, volume 224 of *Progr. Math.*, pages 241–261. Birkhäuser, Basel, 2004.

Roq14. Julien Roques. On generalized hypergeometric equations and mirror maps. *Proc. Amer. Math. Soc.*, 142(9):3153–3167, 2014.

RY10. Mike Roth and Noriko Yui. Mirror symmetry for elliptic curves: the A-model (fermionic) counting. In *Motives, quantum field theory, and pseudodifferential operators*, volume 12 of *Clay Math. Proc.*, pages 245–283. Amer. Math. Soc., Providence, RI, 2010.

Sam09. K. Samol and D. van Straten. Dwork congruences and reflexive polytopes. arXiv:0911.0797, 2009.

Sar12. Peter Sarnak. Notes on thin matrix groups. *Lecture notes at MSRI*, [Supp Item 60], 2012.

Sin14. Sandip Singh. Arithmeticity of Four Hypergeometric Monodromy Groups Associated to CalabiCYau Threefolds. *International Mathematics Research Notices*:rnu217, 2014.

SS97. Jan Saxl and Gary M. Seitz. Subgroups of algebraic groups containing regular unipotent elements. *J. London Math. Soc. (2)*, 55(2):370–386, 1997.

SS03. Elias M. Stein and Rami Shakarchi. *Complex analysis*. Princeton Lectures in Analysis, II. Princeton University Press, Princeton, NJ, 2003.

Str90. Andrew Strominger. Special geometry. *Commun. Math. Phys.*, 133:163–180, 1990.

SV14. Sandip Singh and T. N. Venkataramana. Arithmeticity of certain symplectic hypergeometric groups. *Duke Math. J.*, 163(3):591–617, 2014.

Tak77. Kisao Takeuchi. Arithmetic triangle groups. *J. Math. Soc. Japan*, 29(1):91–106, 1977.

Usu08. Sampei Usui. Generic Torelli theorem for quintic-mirror family. *Proc. Japan Acad. Ser. A Math. Sci.*, 84(8):143–146, 2008.

vdPS03. Marius van der Put and Michael F. Singer. *Galois theory of linear differential equations*, volume 328 of *Grundlehren der Mathematischen Wissenschaften [Fundamental Principles of Mathematical Sciences]*. Springer-Verlag, Berlin, 2003.

vEvS06. Christian van Enckevort and Duco van Straten. Monodromy calculations of fourth order equations of Calabi-Yau type. In *Mirror symmetry. V*, volume 38 of *AMS/IP Stud. Adv. Math.*, pages 539–559. Amer. Math. Soc., Providence, RI, 2006. arXiv:math/0412539v2[math.AG].

Vo02. C. Voisin. *Théorie de Hodge et géométrie algébrique complexe*, Cours Spécialisés, 10. Société Mathématique de France, Paris, 2002.

Voi02. Claire Voisin. *Hodge theory and complex algebraic geometry. I*, volume 76 of *Cambridge Studies in Advanced Mathematics*. Cambridge University Press, Cambridge, 2002. Translated from the French original by Leila Schneps.

Wei67. André Weil. Über die Bestimmung Dirichletscher Reihen durch Funktionalgleichungen. *Math. Ann.*, 168:149–156, 1967.

Wil95. Andrew Wiles. Modular elliptic curves and Fermat's last theorem. *Ann. of Math. (2)*, 141(3):443–551, 1995.

Yui13. Noriko Yui. Modularity of Calabi-Yau varieties: 2011 and beyond. In *Arithmetic and geometry of K3 surfaces and Calabi-Yau threefolds*, volume 67 of *Fields Inst. Commun.*, pages 101–139. Springer, New York, 2013.

YY04. Satoshi Yamaguchi and Shing-Tung Yau. Topological string partition functions as polynomials. *J. High Energy Phys.*, (7):047, 20 pp. (electronic), 2004.

Zag08. Don Zagier. Elliptic modular forms and their applications. In *The 1-2-3 of modular forms*, Universitext, pages 1–103. Springer, Berlin, 2008.

Zud02. V. V. Zudilin. On the integrality of power expansions related to hypergeometric series. *Mat. Zametki*, 71(5):662–676, 2002.

Zud11. V. V. Zudilin. Arithmetic hypergeometric series. *Uspekhi Mat. Nauk*, 66(2(398)):163–216, 2011.

Made in United States
Orlando, FL
04 November 2026

90202228R00111